TOWARD ONE SCIENCE

Toward One Science

THE CONVERGENCE OF TRADITIONS

Paul Snyder

ST. MARTIN'S PRESS, INC.

Copyright © 1978 by St. Martin's Press, Inc.
All Rights Reserved.
Library of Congress Catalog Card Number: 77-86293
Manufactured in the United States of America.
21098
fedcba
For information, write St. Martin's Press, Inc.,
175 Fifth Avenue, New York, N.Y. 10010
cloth ISBN: 0-312-81011-3
paper ISBN: 0-312-81012-1

cover design: Joe Notovitz

Library of Congress Cataloging in Publication Data

Snyder, D   Paul, 1933–
   Toward one science.

   Bibliography: p.
   Includes index.
   1. Science—Philosophy. 2. Science—History.
I. Title.
Q175.S638   501   77-86293
ISBN 0-312-81011-3
ISBN 0-312-81012-1 pbk.

FOR LUTHER

PREFACE

Anyone interested in the sciences, either as a professional or as a spectator, has philosophical views about the nature, methods, and aims of science. My aim in this book is quite frankly to seduce such people into thinking more systematically about their philosophical views concerning science, about their scientific beliefs, and about the consequences of these views and beliefs. Drawing on the work of a number of major writers in philosophy and the organized sciences, I have attempted to lay out a coherent philosophy of science without engaging directly in the critical give-and-take that goes on constantly in the several disciplines I have drawn on. However, I have provided lists of readings at the end of each chapter for readers who wish to explore the antecedents of my view of science or who wish to consider other ways of approaching the important issues.

I think that many of us who try to keep up with what goes on in the sciences have trouble sorting out just what is reasonable to believe about ourselves and the world from what isn't reasonable at all. New developments are happening so rapidly that it is difficult to keep abreast of what the working scientists believe. It is even more difficult to decide whether or not we should share their beliefs at any given time and to determine just what those scientific beliefs have to do with us.

Even more confusing are the beliefs of scientists who directly study human beings. Here the problem for the beginning professional and for the interested non-professional isn't just that the theories and conceptions seem to be changing rapidly, but that there seems to be no mainstream of professional opinion like that found in the physical sciences.

I am not going to suggest that I have found an easy formula for tying together the physical sciences and the human sciences, or for combining the eastern and western conceptions of the self and the world, or for sorting out the many psychological and sociological conceptions that are competing for our respect. I have a more modest aim: I simply want to argue that the tying together can be done, and to suggest a way in which it can be approached.

Philosophy goes through phases, as any discipline does. There are times of analysis and times of synthesis; times of taking things apart to see what makes them tick, and times of trying to see how things can fit together. The analytic times are when professional philosophers tend to spend most of their working time talking to each other; the times of synthesis are when they start messing about in everyone else's melon patch. I think that western philosophy is well into a turn from analysis to synthesis. I hope that this is the case, because trying to figure out how all those melon patches fit together to make a world is where philosophers have all the fun.

This is a book of synthesis in two distinct ways. First, I have tried to put together a coherent philosophical view of science out of a variety of ingredients that I have found attractive over the past few years. Second, I have tried to spell out how this philosophical view of science and many contemporary theories from the several sciences are relevant to each other. This has meant dipping into a number of fields other than my own, drawing on what seem to be standard sources of contemporary opinion, taking what I need for the argument, and moving on to the next point or problem without developing specific points in the detail they might deserve in another sort of book. In some cases, I have taken sides in disagreements that go on in fields other than my own, and to this extent there may be some minor scientific heresies in my lines of argument. I have done my best to ensure that all the *major* heresies are in my own field.

To the readers of this book who have never considered themselves philosophers, I have this to say: If you are convinced that I am dead wrong at any point in the first five chapters, and you are prepared to say why, then you are doing some philosophy of science. That is what this book is really about.

Donald Paul Snyder
Temple University

ACKNOWLEDGMENTS

The major philosophical roots of the view of science presented here are to be found in the work of Karl Popper, Peter Strawson, and Stephen Toulmin, with lesser roots extending to the work of many writers in philosophy and in the sciences.

Many friends, colleagues, and students have contributed to this book over the past several years by helping me to hone the arguments, by supplying examples, information, and references, or by being willing to answer questions about their own fields that have ranged from the naive to the incomprehensible. I am particularly indebted to Charles Reed, who read and criticized early drafts of the sections on psychology and physiology, and to David L. Hull and Robert Weinberg, who made helpful suggestions about the sections on biology and physics.

Among philosophers, I owe particular thanks to Jay Rosenberg for his close reading of the early drafts of this material and his many suggestions for refining and reorganizing it, as well as for his philosophical midwifery. My good friend and colleague Charles E. Dyke has engaged with me philosophically over the past several years concerning the arguments of this book. His criticism, encouragement, and challenging discussion have improved the work at every point of contact. Other colleagues with whom I have grappled on these matters include Monroe Beardsley, Joseph Margolis, and William A. Meroney. Angelos Kolokouris and Thomas Litant were generous in suggesting sources of information.

Finally, I owe personal thanks to my old friend John Gott, historian and bookman, who began my collection of old science and logic books some years ago by giving me the volume by Olmsted which I have used in chapter 5; to George Allen Cook, who first encouraged me to write; and to Grace Stuart for her help in preparing this manuscript.

D.P.S.

CONTENTS

PREFACE   vii

ONE
Some Questions and a Point of View   **1**
Further Readings   *19*

TWO
The Sameness and Difference of People   **20**
The Bare Bones of Natural Selection   *21*
The Natural Selection of Bodily Characteristics   *23*
Down to Cases: How Do People Get Information?   *30*
The Natural Selection of Behavior Patterns   *38*
Down to Cases Again: How Did Languages Develop?   *46*
Further Readings   *52*

THREE
On Making Sense   **54**
Supposing, Believing, and Asserting   *54*
Presuppositions   *56*
Coherent Beliefs and Plausible Hypotheses   *61*
Identifying, Counting, and Classifying   *68*
Describing and Explaining   *70*
Truth, Falsity, and Accurate Descriptions   *73*
Further Readings   *78*

FOUR
Science as an Organized Activity   **79**
Data, Theory, and Nomenclature   *81*
Theoretical Explanation and Prediction   *91*
Further Readings   *99*

FIVE
The Origin of Theses   **101**
The Critical Process at Work: Electrical Theory in 1851   *101*

The Selection of Hypotheses   *110*
The Critical Process Still at Work: Gravitation Theory in 1977   *126*
A Theory About Science itself   *128*
Further Readings   *133*

SIX
Holism and Pluralism as Philosophies of Nature   134
Philosophies of Nature   *135*
Two Clines and Their Histories   *138*
Antiquity   *139*
The Classical Period   *144*
The Middle Ages   *152*
The Revival of Critical Debate   *154*
Holism and Pluralism   *161*
The Limits of the Mechanical World View   *164*
Further Readings   *168*

SEVEN
The Science of Ourselves   169
Admissible Data   *169*
Two Approaches to Helping Ourselves   *176*
Two Approaches to Improving Ourselves   *181*
Convergence   *185*
Further Readings   *187*

EIGHT
One Science   188
What Is Scientific?   *191*
The Criticism of Minimal Hypotheses   *193*
Science and Religion   *197*
The End of Science   *199*
Further Readings   *202*

NOTES   203

BIBLIOGRAPHY   210

INDEX   217

KOSMOS

Who includes diversity and is Nature,
Who is the amplitude of the earth, and the coarseness and sexuality of the
   earth, and the great charity of the earth, and the equilib-
   rium also,
Who has not look'd forth from the windows the eyes for nothing, or whose
   brain held audience with messengers for nothing,
Who contains believers and disbelievers, who is the most majestic lover,
Who holds duly his or her triune proportion of realism, spiritualism, and of
   the aesthetic or intellectual,
Who having consider'd the body finds all its organs and parts good,
Who, out of the theory of the earth and of his or her body understands by
   subtle analogies all other theories,
The theory of a city, a poem, and of the large politics of these States;
Who believes not only in our globe with its sun and moon, but in other
   globes with their suns and moons,
Who, constructing the house of himself or herself, not for a day but for all
   time, sees races, eras, dates, generations,
The past, the future, dwelling there, like space, inseparable together.

from *Leaves of Grass*
WALT WHITMAN

TOWARD ONE SCIENCE

ONE
Some Questions and a Point of View

Beliefs are like pieces in a puzzle. As rational beings, we like to think that our beliefs fit together to form a single coherent picture. But the wild mixture of ideas that confronts us is difficult to fit together coherently. Most of us accept the basic conceptions of western physical science. At the same time, we are confronted with scientific successes that have not come from the western tradition at all, but from knowledge traditions that became firmly established in other ground long before the homogenizing influence of the rapid and detailed communication of ideas. Acupuncture, an oriental technique based on a conception of the human body that is quite different from that of western medicine, produces predictable results. How are we to fit the two conceptions together? Meditative techniques which have been known for centuries in Asia produce results that are substantially the same and apparently more reliable than the new western biofeedback techniques for controlling blood pressure, heart rate, and brain-wave patterns. Can meditation and biofeedback be pieces of the same picture?

The lid seems to have come off in the past few years: some very tough-minded western scientists are beginning to wonder about extrasensory perception, hypnotic healing, biorhythms, and even astrology. At the same time, physicists are talking about particles of matter that aren't particles at all and about processes that violate all the rules of common sense.

We've brought it on ourselves, of course. It is western science that has been responsible for the developments in transportation and

1

communication that make contact among different knowledge traditions possible. And it isn't going to stop with human sources of new ideas: serious scientists are now trying to make linguistic contact with other intelligent species. We may one day be fitting together pieces of that puzzle which originated in the mind of a dolphin. And, as if that weren't enough, scientific associations have already held conferences about communicating with intelligent beings from other planets in other solar systems. Recent space probes have contained plaintive messages, to whom no human knows, identifying us and our planet as the source of the objects.

There is yet another dimension to the puzzle of rational belief. The best theories of any scientific tradition change over a period of time. It becomes difficult to say just what the basic conceptions of physical science are. Not only are pieces of the puzzle coming at us from a variety of directions, they are also constantly changing, and the rate of change is increasing. What physicists had to say ten or fifteen years ago about the ultimate constituents of matter is no longer believed. New items keep coming into the discussion—pi-mesons, fermions, quarks, even "charmed" quarks. In the 1930s we had only protons, electrons, and neutrons. Not too long before that we could content ourselves with indivisible atoms. When can we expect the final word?

I'm convinced that there is a rational point of view that can tie together all these pieces and more into a coherent picture. I'm going to begin by sketching the outlines of what I take to be the major pieces of the puzzle. Doing this will raise questions that will have to be dealt with in the later chapters where I will examine each of the major pieces in more detail and suggest a point of view from which they can be seen to form a single, coherent picture.

Science is something that people do. It is not a particular set of assertions or theories, but a set of activities that may or may not produce organized theories.

The formal theories of the organized sciences are, at any given time, the chief repositories of human knowledge. But scientific activity doesn't stop. Theories change over a period of time. Historically, the theories characteristic of one geographical popula-tion are different from those of others, both in what they explain and in how they explain it. What ties it all together is the activity: human beings explain things as naturally as spiders spin webs, and for much the same reason. Organized into complicated intellectual communities, we explain things in order to survive. As a species, our primary defense and our primary weapon, our primary device for feeding and protecting ourselves, is our intelligence. We solve problems, and we pass along our solutions in organized ways, encoded in the theories of our sciences.

But we can't begin fitting together the scattered pieces by studying theories. That would be like studying the activities of spiders by looking only at the webs that they spin. Fascinating as the webs are, they can't by themselves give us a comprehensive picture of what spiders do, not even of the activity of web spinning, its purpose, and its function in the life of the spider. If we were studying that activity, we would have to begin by getting clear about just what sort of organism a spider is, how and why it came to be that way, why spiders spin webs in the first place, and why different sorts of spider spin different sorts of web.

A similar range of questions confronts us concerning the human activity of doing science. Is there something about the nature of human beings that makes it necessary for us to explain our experiences? Is there a way of looking at the community activity of scientists that can explain why the science that developed in northern Europe is different from the science that developed in China and India? Is it possible that two quite different scientific theories about the same subject matter might both turn out to be right? Can there be such a thing as a global science in, say, psychology, that will cross national and cultural boundaries in the way that contemporary physics does?

I'm going to argue that the answer to all of these questions is *yes*. But the argument isn't brief. To tackle it seriously, I will have to turn to the sciences themselves for major portions of the argument.

Scientific theories explain and predict the kinds of event that fall within their scope. What I am about to develop is a theory *about* science, drawing on the best available theories *of* science to support the line of argument. Like any theory, this one is subject to criticism and revision, and it is always subject to replacement by a better theory. The activity of doing science goes on as a vital part of human experience. Particular theories, including theories *about* theories, come and go. They serve their purpose until they are replaced.

There is something a bit wrenching about trying to fit alien ideas together with familiar ones. It's rather like the shock of maturing. One of the first things we notice about a child as he begins to mature is that his attitudes and behavior gradually shift from totally self-oriented patterns to what seems to be a wider and wider sense of identification with other people. He recognizes that there are other beings like him; that he isn't unique, and that he isn't, for any practical purpose, the center of the universe.

A similar "maturing" has taken place in the recorded history of scientific theories. The earliest recorded explanations of people and the world ran on the assumption that human beings are unique in the universe. Each area of inquiry that we recognize as a distinct field or discipline has gone through terrible conflict when the assumption of human uniqueness has come under question. But the assumption has eventually been dropped in each field. In this sense the disciplines of western science have matured relative to their own subject matter.

In astronomy, for example, we can trace the shift from the view that the earth is the center of the universe to the view that the sun is the center of the universe, and on to the current view that the universe may have no identifiable center at all. The wrench of the maturation was enormous. Only a few hundred years ago people actually died for believing that the universe did not center on mankind. Most of us now are comfortable with the view that there is nothing special or unique about our planet or our position in the universe. We know that there are many stars of the same sort as our sun and, in all probability, many planets of the same sort as our planet, and we know that it's highly likely that intelligent life has developed on some of those planets. The changing theories in astronomy have matured us in that sense.

Biologists, too, have contributed to our maturation. They now view human beings as one species among many somewhat similar species—similar because they all evolved under roughly the same conditions of life on this planet. It has even been suggested that from an intellectual point of view, human beings might not be the most highly developed species on what we like to think of as our own planet. Dolphins have brains that are comparable to ours in size and complexity. Some researchers actually believe that if we can find a way of comparing human intelligence with dolphin intelligence, we might come out second best.

Even psychology, perhaps the most human-centered discipline of all, has moved away from the assumption that human beings are unique. If we share a genetic history with other species, then we can learn something about human thinking and behavior by studying those other species. Complex human behavior should be echoed, in simpler form, in other species that aren't as complicated neurologically as we are. So the study of how rats learn to solve problems and of how chimpanzees can learn rudimentary symbolic languages becomes relevant to human psychology.

What I suggest in this book is that the shift away from egocentricity involves at least one more important step before we can claim to have a mature view of scientific knowledge. The attitudes of many western scientists and philosophers toward science itself reflect the belief that the western tradition of science is somehow uniquely correct. I will argue that western science itself indicates that these attitudes are untenable, that the western approach to knowledge is neither more nor less correct than that of other traditions and that, even if we ultimately succeed in integrating the best knowledge of all human traditions, human knowledge is not the only possible sort of knowledge.

An attempt to fit together beliefs that originate in disparate knowledge traditions cannot begin with the surface statements that represent the present state of those traditions. Even to integrate statements from two different fields within western science entails digging below the surface. A theory about human science has to begin with the questions of why human beings do science at all and why they do it in a particular way at a given time and place.

Right away, these questions raise more questions: How do theories change within a scientific tradition? Why do they change, and who decides which changes are going to stick? How does a scientific tradition develop? What is there about the "nature" of human beings that leads ultimately to doing science? To get at questions like these, we have to start at square one and talk about how any species— and the human species in particular—comes to have the characteristics it has.

It is comfortable to think that human beings can shape and control the world. Civilization, from its very beginnings, rests on the ability to control and direct nature. Century by century, we have learned to manipulate natural forces and events to our advantage. We are able to meet conditions as they change, and we are able to flourish on every part of the planet. We are the most flexible species on earth.

But if you take your biology seriously, the opposite is also true: the world has shaped us. Century by century, the facts of life on this planet have made us into the kind of creatures we are. The world determines what we need to do in order to flourish, and beyond that, the characteristics of this planet have shaped the physical equipment we have to do it with. The shaping takes place in the process that Charles Darwin called natural selection.

One tends to think of natural selection, with its strong emphasis on survival, as having to do with the "hardware" of organisms: teeth and claws, legs and spinal columns, lungs and opposable thumbs. But of equal importance to an organism are its sensory equipment and the nervous system that guides its activities in response to the stimuli that the sensory equipment picks up. The most significant development for human beings is not a natural set of physical weapons like tusks or claws, but rather the physical apparatus that makes thought and language possible. We are the species that lives by its wits.

Coupled with the development of sensory and neural apparatus is the development of techniques for using that apparatus in systematic ways to deal with the environment, techniques that are made possible by the physical characteristics that the species has and that at the same time are constrained by those physical characteristics.

The development of techniques—typical patterns of behavior, if you like—is evolutionary in the same way that the development of bodily characteristics is. This matter is somewhat problematic for biologists, psychologists, and physiologists at the present time. Although the way in which a particular behavioral trait is passed along is not always clear, it is important to recognize that techniques do develop as adaptive behavior and that these techniques are passed along as characteristic ways of solving problems.

Spiders spin webs to catch their food. Wolves travel in highly organized packs. Birds migrate to feeding and nesting grounds. Beavers build dams. Human beings explain things. Singly and in organized groups, they develop elaborate theories to explain why events happen as they do. By explaining past events, human beings are able to manipulate future events to their advantage, both in the obvious cases of finding food and shelter, of breeding and defense, and in the more complicated cases of traveling over long distances, forming communities, and developing means to support such communities. The more complicated patterns of behavior are in large measure solutions to problems that arise at least in part from earlier solutions to more basic problems.

Here, in broad outline, is one piece of the puzzle concerning human science: certain sorts of activity, which we identify in ourselves as conceptual activity, are best understood as part of the naturally selected techniques by means of which our species manages to survive. Now we are still a long way from talking about the formal development of scientific theory, let alone talking about integrating theories from scattered knowledge traditions.

There are still more questions. How are workable solutions to problems sorted out from unworkable ones? How can natural selection explain the solving of problems that affect whole populations? And all of this precedes our getting at the main questions I'm after here: What counts as a scientific approach to a problem as distinct from an unscientific one? How can we combine the most successful scientific explanations available to us into one coherent picture of reality?

If we are going to talk seriously about how human beings explain their experiences, we will need to consider the sorts of experience we can have. Human beings experience the world with a particular sensory apparatus: we have eyes that are sensitive to radiation in a narrow band of the electromagnetic spectrum and other senses that are stimulated within given ranges. We don't "see" all wavelengths of reflected radiation or "hear" all frequencies of atmospheric vibration. Taken together, the human sensory ranges constitute a particular point of view from which we experience the world around us.

Organisms have the particular sense organs that they do for reasons that a biologist can explain by making reference to the peculiarities of the planet on which those organisms evolved. The size of the planet, its distance from its sun, the composition of its atmosphere, all serve to determine the character of the most readily available "information"—sensory stimuli that can be of use in detecting food and danger. It is no accident that most life-forms on this planet have sensory apparatus that react to roughly the same sorts of stimuli within roughly the same ranges. The differences in sensory ability from one species to another are easily enough explained in terms of the relatively local conditions under which particular species developed and the special local needs that made, for example, the chemical senses of taste and smell more important than the visual sense to dogs, and the visual sense more important to human beings.

Our brain and neural system developed as they did, step by step, because each step in the development enhanced our ability to interpret the available stimuli and act on them in order to meet our needs. The facts of human life on this planet could have been different from what they have been. If the available "raw information" had been different or if what we needed to know had been different, our means of gathering and processing information would have been different too.

Now what does this suggest about nonhuman intelligence? Suppose we can make sense of the claim that dolphins are as intelligent as we are. Is it at all likely that they "think" the way we do? Perhaps before long we will be able to find out whether or not the rapid, complex sounds that dolphins make constitute a language that is as sophisticated as human language, and, if it is such a language, we should eventually be able to get some idea of its structure and organization.

But consider this: The interpretation of visual stimuli employs a larger portion of the human brain than does the interpretation of stimuli from the other senses, and it seems to be the case that the visual sense provides the organizational scheme into which information from the other senses is fitted. With dolphins, the situation appears to be different. Their "organizing" sensory system, by all indications, is their sonic echoing apparatus (which, with a similar sense that bats have, gave us the idea for radar and sonar). What kind of natural language could be based on such a sense? What sort of object would they converse about? What sort of properties would they discern those objects to have?

There is one more question before we set this subject aside. If the characteristics of available sensory stimuli and indeed the characteristics of human intellectual activity are structured by the particular facts of life on this planet, what sort of intelligence could we expect to find in members of species that developed on other planets where the facts of life are likely to have been quite different?

I will return to these questions several times, but they have to be kept on the margin of the main line of discussion. We just don't know enough yet to develop reasonable answers to them. But they are similar to a group of questions that we must ask about human beings: To what extent have different circumstances and different needs influenced the sort of language developed by human populations? To what extent do circumstances, needs, *and* languages affect the problems that are systematically addressed by a human community and the knowledge that develops out of the solutions to those problems?

Even if we restrict our attention to a single species and to the particular equipment it has for dealing with its needs, we find that solutions to problems differ from one physically similar population to another. Any given set of physical characteristics makes possible a wide variation in the techniques that a species can develop. Some species of shore birds differ from flock to flock and sometimes even within a flock in their characteristic techniques for opening shellfish. Some species of spiders show regional differences in the style of web that they spin, again without any discernible difference in the spiders.

Within the human species there are populations that differ with respect to the character of the languages in which they encode information and that have differed, from one major geographical region to another, in the kinds of information that they tend to systematize and pass along to their offspring. It is no accident that different sorts of language and different bodies of organized knowledge developed in different parts of the world. Local conditions of terrain, climate, availability of water, food supply, presence or absence of predators have all influenced the sort of problem that must be addressed first as well as the technique that will be effective in dealing with a given problem.

Now that first problem is important. We can't know precisely, of course, the first problems for which our remote ancestors in a given part of the world developed solutions, but we do know that the established techniques—the accumulated body of successful solutions to problems—serve at any given time to influence the sort of approach that will be taken to a new problem. Over a period of time a characteristic style of conceptualization or local idiom will develop.

Without major physical differences in sensory, neural, or other apparatus, for that matter, human beings have learned to deal with the world conceptually and linguistically in several distinct ways. Each geographically separated population has preserved its successful solutions to problems and has passed them along to successive generations. It is only very recently in human history that these separated populations have come in close contact with each other. The enormous successes of western physical science have been responsible in large measure for the contact. And because of this, the western physical sciences have come to be the primary models for what a science should be. The demand has been made in various ways for psychology, sociology, and even biology to be "more like" physics and chemistry.

More questions, immediately. What makes a theory successful? What does a scientific theory *do* for us? Why should we expect all successful theories to resemble each other?

Because knowledge traditions that have developed elsewhere in the world differ in so many obvious respects from the conceptions of western physical science, they are viewed with suspicion by many western scientists. What I hope to convince you is that science is one kind of activity that can be carried out in many different styles; that the style of western physical science is already beginning to overlap with the style of eastern thought; and that the differences between the characteristic approaches to knowledge developed in geographically separated populations are differences that reflect the ways in which those populations learned to deal with the problems that confronted them.

You can't learn to do what you lack the equipment for, however. There are conceptual characteristics of human beings that reflect, so far as we can tell, the boundaries of what our sensory and neural equipment make possible.

What exactly are the limits of possible variation in human thought? We will have to turn to psychology and philosophy to trace the contours of the next piece of the puzzle. The physiology associated with conceptual activity is still something of a mystery, and at present the best evidence we have about how people think is to be found in what people say and in what they are able to do conceptually.

Can you imagine a four-dimensional physical object? "Imagine" must be taken very literally here, as in imaging or picturing. The answer is probably *no*, and it is probably the same for every human being, even though we can "do the math" for four-dimensional objects. Geometry that deals with objects of more than three dimensions has been around for a long time. Just what kind of limitation are we talking about? We certainly don't *see* objects in four-dimensional space (which is different from saying that there are none); our visual imagination simply doesn't seem to allow us to have images of four-dimensional physical solids.

When physicists talk of multi-dimensional objects and spaces or of particles that don't have spatial dimension, position, velocity, or even (in the case of the photon) mass, they are talking about objects that we can describe mathematically but that we cannot experience with our senses or visual imagination. But why not? There isn't anything in the math itself that says we can't. Physicists talk freely about gravity as three-dimensional space *curving* in a fourth dimension, and they seem to be able to make perfect sense of it. They can even argue among themselves concerning the accuracy of each other's descriptions of objects in four-dimensional space.

Whatever we can say about four-dimensional objects, they are not objects of common sense. This is a matter I will return to later, but let me give a minimal characterization here of what I take to be common sense experience of objects.

First of all, the physical objects we experience exist in three-dimensional space. And, second, events in which those objects figure happen in an irreversible sequence in time. Here again we can talk of reversible time sequences in our physical theory, and we can construct fictional accounts of travel backward and forward in time, but these things are really beyond common sense experience. Finally, it seems to be a precondition of our making sense of events that every event must have a cause. Later I will discuss just why it is that these three basic characteristics of human experience, along with several other characteristics, seem to be universal. For now, I will take space, time, and causality as the minimal list of features common to all our sensory experience.

Let's grant, even, that all human beings have the same basic experience of common sense physical objects. The objects have the sensible properties of color, texture, smell, taste, and sound. They occupy three spatial dimensions, they endure for at least a period of time, and they are solid—at least solid enough for us to discern their boundaries as discrete objects. Might there be a uniquely correct way of describing a system of such objects that we could take as a beginning point for human knowledge?

Consider the walls of the room you are in right now as the boundaries of a physical system. (If you are reading outdoors, construct some arbitrary boundaries around yourself, marked off by trees or rocks, so that you have a room-sized space defined.) Having defined the physical space, then, we want to try for a neutral and correct description of the physical objects in the room.

But *how many* physical objects are there in the room?

That sounds like a silly question, doesn't it? Nobody would really ask it outside a philosophy classroom. On the face of it, it might seem that finding the correct answer would simply involve a few tedious hours of counting the objects.

But it's not that simple. Embarrassing problems come up almost immediately. Should you count your loose-leaf notebook as one object, or should you count each page separately? What about tables and chairs? Is the chair one object or twelve pieces of wood? One ring of keys, or five keys and a ring? Do you dismantle the lamps and count each part separately, or just count the lamps? Each of these choices could reasonably go either way, and each choice affects the outcome of the count. So there is more than one "correct" number of objects in the room. It depends upon what you decide to take as one physical object.

There might be reasonable arguments in some cases about the best way to count objects, but the problem comes down to this. The best way to count depends upon why you are counting the objects in the first place. A furniture dealer would count the chair as one object; a cabinetmaker might count the twelve pieces of wood separately. A zealous physicist might give you a fair estimate of the number of molecules, or even atoms, in the chair. What is the correct number of objects? Well, who is counting, and why?

Once you have decided on a *way* to count, you can talk about right and wrong answers to the question. But "physical object" and "thing" don't give you a way of counting. You need to specify more about what is to be taken as a single object, and there is no uniquely correct way to do this.

Now move the question up a notch: How many different *kinds* of object are there in the room? The same sort of difficulty arises. There are indefinitely many ways of sorting out objects into kinds. How many colors do the objects have? Same problem. You have to make a decision about *how* to sort things out into kinds, if only the decision to use a particular scheme of classification that you get from someone else. You have to establish or adopt a particular way to draw the boundaries between colors. There is more than one reasonable way of doing each of these things within the limits of what we *can* do.

This is not to say that we can never differentiate between correct or incorrect counts or descriptions, only that there is no *uniquely* correct way of counting, classifying, or describing a system of objects. You might, in a given case, argue that one sort of description of a physical system is better than another for a given purpose, but there are many different sorts of description, each of which might be appropriate in some context or other and each of which determines a different "correct" count, classification, or description.

Descriptions are never neutral. There is no uniquely correct way of describing physical systems or events, and there is no uniquely correct way of explaining them either, since an explanation of why a thing has a given set of characteristics must depend upon what we take those characteristics to be. We describe and explain from a point of view, but there is no point of view that is common to all human beings in the way that the framework of space and time seems to be. So, even if there is something peculiarly correct about saying that the objects we experience exist in three-dimensional space, there doesn't seem to be any peculiarly correct way of describing those objects. We can't even speak here of a clear-cut "western" style or "eastern" style of description. So many reasonable variations have equal claim to accuracy that within a given local idiom we still can't sort out one kind of description that is uniquely correct.

Now this raises a question that bothers some scientists a great deal. If there are a number of more or less arbitrary choices to be made before we can talk about the correct description, or even the correct number of objects in a given space, is it possible to claim objectivity for any of our judgments about the world? How can we talk about one science, or even one correct scientific theory, when there are indefinitely many ways of describing a physical system as ordinary as the objects in a single room?

The question of objectivity will recur several times as this theory about theories develops. For a start, we should sort out claims for objectivity from claims for neutrality. We cannot make judgments without a point of view, and there is no point of view that is absolutely neutral or, in some final sense, correct. That much has emerged from the questions about the numbers, kinds, and descriptions of objects in the room. The framework of space, time, and causality is basic to all human thought, the most sophisticated and the most commonplace, but it is only the beginning of the characterization of a context in which we can make sense of descriptions of the world and assess their accuracy.

A point of view can be characterized further by what I will call *presuppositions*. These can be expressed as assertions of a particular sort, about what kinds of thing there are, about what specific individual things exist, about what properties things of a given kind can have. Presuppositions define the contexts in which intelligent behavior and discourse can take place and in which solutions to problems, including the theories of the organized sciences, are passed along from one generation to the next.

More important, a context of presuppositions—a point of view— must be understood before we can decide whether or not a given description is accurate. Aside from what I will call the presuppositions of common sense—space, time, and causality, for a start— other presuppositions are subject to wide variation within what human beings are capable of doing linguistically and conceptually. They are determined by factors other than our sheer physical makeup: language, culture, local idiom, what we have learned to distinguish, what we need in a given situation. We can replace these presuppositions by others with some effort at relearning. Still other presuppositions can be replaced easily and define idioms or "manners of speaking" that we shift into or out of as day-to-day situations demand.

Quarks are held together by gluons. That's what the physicists say. Are they correct? Does the assertion make any sense? How can we tell?

First, we would have to know something about quarks and gluons and about how the expression "held together" is to be understood in the context. Then, we would need to determine whether or not the assertion *can* be said to be correct before we can decide whether or not it *is* correct.

In Chapter 3, I will go into some detail about what is involved in making such determinations. But let me illustrate here part of what is at stake. Suppose I were to tell you that the drugstore at the corner of Broad and Norris in Philadelphia was robbed last night. Am I telling the truth or not? If you know the intersection, you know that there is no drugstore there. But you can't properly say that my assertion about the drugstore's being robbed is either true or false. To say that it is true commits you to saying that there is such a drugstore and that it was robbed. To say that it is false commits you to asserting that the drugstore at the corner of Broad and Norris *wasn't* robbed last night, which still leaves you talking about a nonexistent drugstore. The appropriate way to treat my assertion is to reject it altogether: "That isn't a drugstore, it's a university," or simply, "There is no such drugstore." In asserting that a given object has such-and-such a property or that it figured in such-and-such an event, we normally *presuppose* that the object exists. If it does not exist, the assertion is *improper*.

The assertion about the robbery isn't nonsense, of course. We know perfectly well what it would take to make it proper. There would have to be such a drugstore. If there were, then it would be a straightforward matter to determine whether or not it had indeed been robbed last night.

Now, let's take another kind of case. Suppose I were to say, "My typewriter isn't happy today." The difficulty here isn't that I am talking about something that doesn't exist. I do have a typewriter. Perhaps, with a little imagination, you might make some sense of what I have said as a figure of speech: a key is sticking, it needs cleaning or a new ribbon, something like that. But, as an assertion to be taken literally, we can't allow it. Certainly it isn't true, and if we say that it is false we are saying that my typewriter *is* happy today, which is just as problematic as the assertion that it isn't. We have another improper assertion on our hands, because a typewriter just isn't the kind of thing that can be either happy or not happy. So there is another sort of presupposition that must be taken in order for assertions to be proper: Whatever we are talking about must be something that can reasonably be said to have, or fail to have, the property that we attribute to it.

One more example, a little closer to our interests here. Suppose I were to suggest the following line of reasoning. The temperature of your body is approximately 98.6° Fahrenheit. Let's be content with saying that it is somewhere above 95° F. It would seem reasonable, then, to say that the average temperature of your internal organs is somewhat above 95° F. No problem so far.

But as we move downward in scale and size, our attribution of an average temperature runs into conceptual trouble. We can reasonably speak, perhaps, about the average temperature of the cells which make up our internal organs, but not of the average temperature of the molecules which comprise the cells, and surely not of the atoms and subatomic particles which constitute the molecules. The claim that a given molecule, atom, or electron has a temperature somewhat above 95° fails, not because the temperature is below

95°, and not because you can't find a place to stick the thermometer, but because molecules, atoms, and electrons don't *have* properties of temperature, although some of their properties are of course related to the temperatures of the systems in which they exist.

It is easy enough with made-up examples to track down just how an assertion misfires. In more complicated cases, where one group of people makes an assertion and another group claims that it doesn't make sense, it is not so easy.

Presuppositions can be understood as rules for making sense. Some of the rules will change over a period of time. If you think in terms of the classical conceptions of Newton's physics, which are very much like what I have called the presuppositions of common sense, you can see perfectly well that the current theories about subatomic events just don't make sense in Newtonian terms.

This isn't a particularly new or startling point. Most of us are familiar with the fact that our view of physical reality has changed several times in the physical theory of this century. The dramatic impact of Einstein's work and that of his successors has led some people to think that truth and falsity are matters that we can no longer discuss intelligently; that a given assertion can be "true for me" whether or not it makes a grain of sense to anyone else. This just won't do, of course. Such a situation would make group solutions to common problems impossible. It would make science impossible on both local and global scales. So there are still more questions to be dealt with. If our assertions can be assessed only within a well-understood context, how can we choose the best context in which to frame our scientific theories? How can we ever compare two theories that are constructed within different contexts of presupposition?

If a Freudian psychologist starts talking about, say, Richard Nixon's superego, a behavioral psychologist is likely to claim that the Freudian isn't making sense. Another Freudian might disagree with the first about some of his assertions, but there wouldn't be a question between the two Freudians about whether or not the assertions were proper. Now who is right here? The Freudians or the behaviorists? In such cases, we sometimes speak of the Freudian idiom or the behaviorist idiom. Do we just have to note the difference between the two idioms and let it go at that, or can we make some reasonable connection between the two? The same questions arise when we compare two theories about the same subject matter that are separated by time or by geography.

I suggested earlier that human beings need to explain, that the activity of theorizing about why things happen is as important to us as the activity of web spinning is to a spider. This isn't to say that fully developed formal theories spring up in some spontaneous way—indeed, they do not, and I will describe in Chapter 5 how they do develop. What I am suggesting is that the kind of activity that has led to the formal theories of western science is something that all people do.

If you should hear a loud crashing noise just outside your room, you might decide that it has nothing to do with you and simply ignore it. You might, more likely, open the door to find out what happened, or you might prudently pause for a minute and consider what could have caused the noise before opening the door to investigate. What you *wouldn't* do is suppose that there was no reason at all for the noise. Things don't just happen in our experience. Every event is caused. When an event seems relevant to us and our well-being, we want to know what the causes are. When they aren't immediately apparent, we develop hypotheses—suppositions about what the causes might be—and we sort out the plausible hypotheses from the implausible ones before acting on any of them. Here again, there are questions that we will have to try to answer: How do we decide which hypotheses are plausible and which are implausible? How do we choose among several plausible hypotheses?

Beyond explaining why particular events happened, we try to describe reality in general terms. But, as we saw earlier in trying to count and describe the contents of a single room, there are many different ways of describing things. Is it at all consistent to say that there is one reality, but many different ways of describing it? Can there be more than one *correct* way of describing reality?

Several years ago, a physicist, a psychologist, and a philosopher were jointly teaching a course called "Scientific Knowledge" to a group of fifty university freshmen.* On a particular day, the physicist was lecturing about light and color. He described how objects reflect light of different wavelengths and showed how given wavelengths of light are identified as given colors. He demonstrated how objects which appear to be one color under a particular kind of illumination appear to be quite a different color under other kinds of illumination.

*The physicist was Robert Weinberg and the psychologist was Charles Reed, both of Temple University.

Finally, the physicist pointed to a red shirt and asked, "What color is it *really*, all by itself? If you take the red shirt and put it in a drawer, what color is it?" The answer he wanted and got was "No color." No color, because color is a property of light. When the shirt is in a drawer, it isn't reflecting any light at all, let alone that portion of the visual spectrum identified as red light.

The philosopher objected immediately. Did the physicist seriously believe that all the shirts in his drawer were no-colored shirts; that they *became* red or blue or whatever only when the drawer was opened in a lighted room? Suppose he called a department store and ordered a red shirt. If it was delivered in a sealed package, would he send it back to the store unopened with the complaint that the shirt wasn't red? Wouldn't that be a little odd?

Surely, the philosopher argued, the *shirt* was really red, and all the physicist's talk of reflected light simply explained why red things appear red under normal illumination. Color is a property of objects, he insisted, and our way of telling what color an object *really* is consists of viewing it under normal light.

Now the psychologist objected. Reflected electromagnetic radiation of given wavelengths is *just* reflected radiation, he said; it isn't color. It may be a fact that certain dyes and pigments reflect radiation of given wavelengths more than others, but you can't talk of *color* until you bring the human visual system into the discussion. Color properties are a function of the objects, the radiation they reflect, and the way in which that radiation is apprehended and processed by the human visual system. So color is neither a property of the light nor of the object. It is *really* a property of the complex perceptual experience.

Now who is right here? What color is the shirt *really*? What is color *really*? What is color a property of? Objects? Light? Experience? Could all three of them be right?

Every day, we literally stake our lives on the adequacy of our hypotheses and the theories that develop from them. In most cases, our theories are successful in guiding our actions safely. Whether or not a theory is successful in this sense does not depend upon the particular field, or even the particular tradition, in which it developed and became established. Our beliefs have far-reaching roots. A coherent pattern of beliefs can—perhaps must, at this time in human history—contain elements that originated in widely divergent traditions of knowledge. But how do we integrate such disparate ingredients?

This is a long list of questions, and each must be dealt with if we are to construct the coherent picture of rational beliefs that I spoke about earlier. To get at the questions in a reasonable way, we will have to begin with a general account of how human beings come to have the characteristics they do and how some of those characteristics have led to systems of organized knowledge in every human civilization. We will have to consider just how it is that we are able to make assertions about the world, and how we can choose between conflicting assertions. Beyond this, there are questions about how we can sort out the objects of our experience in systematic ways, develop theories to explain the characteristics of those objects, and predict what we can expect from those objects in new situations.

Finally, we will have to develop a general account of what it is to do science, whereby Aristotle's physics, Newton's physics, contemporary physics, eastern and western psychology, and other bodies of theory all emerge as products of the same kind of activity, to explain how they came to be so different from each other, and to suggest how they can be brought to bear on each other in a coherent way.

And that is what I propose to do in the chapters that follow.

FURTHER READINGS

Most of the topics raised in this chapter will be dealt with in more detail in subsequent chapters, and recommended reading will be given at those places.

For an interesting discussion of attempts to visualize in four spatial dimensions, see Ludwig Eckhart, *Four-Dimensional Space* (1968).

TWO
The Sameness and Difference of People

If we are going to be serious about trying to answer the many questions introduced in Chapter 1, we have first to attack some larger and more difficult questions: How much alike are human beings intellectually? How much do human beings differ from each other in the characteristic ways in which they understand the world? And in both cases, Why?

To speak of sameness and difference, we have to speak about change. A cluster of the questions already raised can be tied together as a series of questions about change:

Why do populations of living organisms change in their physical characteristics over a period of time?

Why do populations of living beings change in their habits over a period of time?

Why do human languages change over a period of time?

I'm not going to suggest that these are all the same question, only that they are the same kind of question at a reasonable level of abstraction. I do think that they have the same kind of answer and that the answers are linked to each other, but that remains to be argued. It isn't established simply by noting a similarity among the questions.

Each of the questions is about change over a period of time within a complicated system of interrelated and similar components. The structure, in each case, is in a state of constant activity. They are all dynamic systems rather than static ones.

The first question has a famous and well-respected answer. Charles Darwin's theory of the origin of species seeks to explain how variations in bodily traits become established in populations of living organisms.[1] But before we look at the contemporary version of Darwin's answer to the first question, I want first to look at the *kind* of answer it is by laying out Darwin's theory in terms that are relatively neutral so far as the subject matter is concerned.

The Bare Bones of Natural Selection

Darwin's problem, like that of his predecessors, was this: How can changes in the characteristics of complicated living systems be made intelligible?

If we freeze the action in such a system by taking a photograph, go away for a time, and then return to take a second photograph, we will have recorded at least two kinds of change. First, the individual components of the system will have shifted around in their relationship to each other. But more important, if we focus our attention on the original components of the system, every one of them will have changed slightly in its characteristics in the time between the first photograph and the second.

Are the changes random, like the shifting patterns of a kaleidoscope? That seems unlikely, because the way in which each individual component has changed seems to be related to the way in which every other component has changed.

There is a story to be told, then, of how the changes took place between the time we took the first photograph and the time we took the second. Darwin insisted that the story must have a plot. An easy and obvious plot would introduce some outside interference in the living system that simply shifted the components around and changed their characteristics by sheer brute force. But that doesn't make for an interesting plot or an instructive story, because it doesn't tell us why the changes took place from the point of view of the system itself. Moreover, it involves introducing an implausible, invisible element into the story for no apparent purpose except to move the story along.

So Darwin focused his attention on this question: How does a *single* change in *one* component of a living system come about? The story now has a protagonist—that one component. A second question follows: How does the change become established in the overall system? Now there is a line on which to develop a plot.

Here is the kind of plot Darwin came up with, stripped down to its bare schematic bones. The first photograph we take gives the setting of the story. The action is frozen for an instant, and a state of *equilibrium* is observed; the components of the system stand in some particular relationship to each other. But the equilibrium is *problematic*. We have stopped the constantly shifting action artificially, just to get the setting for the story.

Now the action starts again. An *innovation* occurs. One component is slightly different from what it was when we established the setting. Immediately, the altered component faces an *internal challenge*: Is it viable in its own right? Can it persist at all in this setting? If so, the component faces a second, *external challenge*: Is the altered component compatible with the other components which are present in the setting at this time? If the altered component meets both the internal and the external challenges, it is established at some *location* in the overall setting in relation to the rest of the components, where it persists for a time.

Now we take a second photograph, to stop the action again. A *new problematic equilibrium* is observed, and we can see how this one change has affected the system as a whole. (See Figure 2-1.)

The plot outline is evolutionary in that it tracks the development of a system from one state to another, altered, state. But the important thing about the process I have been describing is that it involves *selection* of those altered components which appear in the setting. The selection of components which will eventually find a location in the system is a matter of the relationship between the new component and the overall system itself, at the third and fourth steps of the plot.

The term "selection," as used by Darwin, should not suggest that some outside agency is involved. It is the relationship between the altered components and the system itself that determines viability, and the relationship between the new component and the other existing components that determines compatibility. The selective process takes place *within* the overall setting.

There are two important questions left unanswered in the plot outline of selection: How does the system originate in the first place? Why does a given alteration occur? But these questions are extraneous to the selective process itself. The plot need not begin at the dawn of time. It can start at any arbitrarily selected time. What is important to the basic evolutionary plot is the way in which the

Figure 2-1 • The Bare Bones of Natural Selection

The six stages isolated here are not Darwin's, of course, but they can be distinguished in his overall account of evolution. The schema here is a direct descendant of Karl Popper's schema of the method of trial and error-elimination.[2]

structure itself selects, and is altered by, those innovations which do occur. This becomes clearer as we examine the plot in a more fleshed-out form.

The Natural Selection of Bodily Characteristics

Now, to Darwin's explanation of organic change. Darwin did not attempt to tell the full story of the development of each species and variety of organism that exists, and no biologist today would attempt to give you the full story, filling in every step from the most simple microscopic organisms to living beings as complex as human beings and dolphins. What Darwin gave science is a plot for many stories, in order to explain how those steps could have taken place. Darwin's theory has been changed somewhat since he introduced it in the middle of the nineteenth century, and there have been a number of disagreements and criticisms. But the scientific community is in general agreement on Darwin's central thesis: Natural

Figure 2-2 • The Natural Selection of Bodily Characteristics

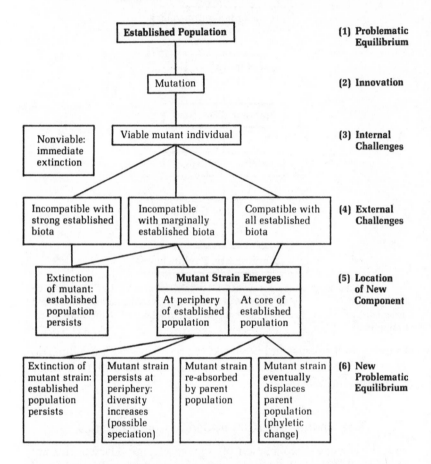

selection is what controls the bodily development of organisms over a period of generations.*

To get at how Darwin explained bodily change, I want to explore, as we did in the "bare bones" version of the plot, the possible sequence of events surrounding a single innovation. The stages of Darwin's plot for the stories of natural selection are shown in Figure 2–2. The plot begins just before an innovation occurs.

*There are some characteristics of organisms that are difficult to account for along Darwinian lines.[3] There are questions as well whether or not a theory as general as Darwin's can have empirical significance.[4] There has also been disagreement about

[1] *Problematic Equilibrium.* We begin with an established population with reasonably discernible boundaries. The unit recognized by evolutionary taxonomy is the *species*, understood now as a lineage of populations evolving separately from others and able to reproduce itself by producing fertile offspring. Bear in mind that not all members of a given species are exactly alike. There can be enormous diversity in the physical characteristics of individuals within a single species. Great Danes and Chihuahuas, for example, are members of the same species, along with the other breeds of domestic dogs, and all the mixed breeds as well.

Each identifiable feature of an existing population—its mechanisms for nourishment, breeding, defense, and flight from predators, as well as other bodily characteristics—may be viewed as a solution, or a complex set of solutions, to the constant problem situation: survival and reproduction in a continually changing environment. Any given innovation thus takes place against a background of established bodily patterns. New organic structures arise within a context of existing structures. These organic structures exist in a larger system of *biota*—all the plants and animals of a region— which, taken together with the gross features of the environment, constitute the larger setting in which a given population lives in a state of problematic equilibrium.

[2] *Innovation.* Living organisms don't reproduce exact replicas of themselves. They produce offspring which differ from the parents in minor ways. Most simple organisms produce offspring in enormous quantities—so enormous that if they all lived to reproduce at the same rate the earth would very quickly be unable to hold them all. The number of small variations among the individuals in a single

---

the rate of evolutionary change. Darwin and his followers held that the rate was fairly constant and very slow, and that evolutionary change almost always involved the replacement of an entire population of organisms over a long period of time by an altered population of descendants (phyletic change). The sparsity of fossil records to supply the links between one known species and another is attributed to the destruction of fossil remains by geological upheaval over the centuries.

Some of Darwin's contemporaries, and many present-day biologists, believe that the rate of evolutionary change has been very uneven indeed, even cataclysmic at times, with significant change taking place in far fewer generations than Darwin would have thought possible.[5] They point out that there is much greater diversity within a population than Darwin had supposed: greater variation in minor characteristics that can lead to the rapid selection of one segment of a population over another in times of natural crisis such as climatic change, geological disturbance or drought.

generation is likely to be quite large. The particular kind of variation that is of interest here is called *mutation*.

What distinguishes mutation from other sorts of variation in individual offspring is that mutation is the result of change in the genetic material of the offspring. This means that the variation can be passed along to future generations. Other variations in individuals, which may have to do with drastic change in the parental diet, accidents of birth, or mischance during gestation, are not mutations, so long as the genetic material is not altered.

Effectively, mutation occurs in a random way. Exactly why this is the case need not concern us here. What is important is that mutations do happen, producing many minor differences between parent and offspring that can be passed along to further generations in the genetic material.*

[3] *Internal Challenges.*   An individual that is genetically different from its parent stock is called a *mutant*. Many differences from one generation to the next are so slight that the mutant's immediate prospects for survival are nearly identical to those of its parents. The variations make no difference in the offspring's ability to nourish or defend itself or to adapt to changes in the living situation.

Other differences between parent and offspring are significant. Some carry a slight advantage of one sort or another over the characteristics of the parent stock. Others are crippling, and the offspring die very quickly. The mutants may be hampered in their ability to breathe or take in nourishment, for example. They simply cannot sustain life on their own in the environment, and their altered genetic material dies with them. It never gets passed along to successive generations.[6]

If the individual mutant does prove to be viable—that is, if the genetic change is either neutral or advantageous—it becomes part of the total breeding population. It can then produce offspring of its

---

*The mechanics of mutation were unknown to Darwin. Despite a long tradition in the study of genetics, it is only recently that molecular biology has come up with what promises to be an explanation of mutation in terms of the genetic molecules DNA and RNA, which are complex enough to be altered in minor but significant ways by chemical, physical, or radioactive agencies present in the natural situation. Any single mutation from one generation to the next is likely to be very minor indeed. The alteration of the preexisting genetic structure will be barely noticeable, unless there has occurred a particularly unusual set of circumstances, such as high doses of radiation, at a crucial time in the reproductive cycle. How often and how rapidly such alteration occurs varies considerably.

own, some of which will exhibit yet further variations, and so on, interminably.

[4] *External Challenges.*   We have followed the mutant to the point where it can stand on its own. Now the question arises whether or not the new genetic material is going to persist in a mutant strain—a subpopulation of genetically altered individuals descended from the individual mutant. This depends upon where the mutant stands in relation to the total environment. What difference does its difference *make* in its ability to get around the local terrain or endure the climate? How does it fit in with the *biota*—the plants and animals of the region, including its own parent population?

Isolating the outcome of a single mutation is impossible in a complex, constantly changing life-system. But we can sort out three possible situations.

First, the mutant individual may be incompatible with well-established biota that destroy it and thereby destroy its new genetic material. It may be more vulnerable to predators, parasites, or disease-bearing organisms or perhaps allergic to or poisoned by the plant life of the region. Whatever the particular case, if the mutant is incompatible with well-established local life-forms, it will almost certainly not establish a mutant strain of genetically altered offspring.

Second, the mutant may be incompatible with local biota that are only marginally established—not doing too well themselves. Conflict. The mutant might or might not establish a strain of offspring. One of the two—the mutant strain or the local life-forms with which it is incompatible—will eventually displace the other.

Finally, the mutant may be compatible with all the established local biota to at least a tolerable degree; its genetic material will likely be passed along to successive generations, and a mutant strain will be established as a subpopulation within the total population, which by now includes other new strains as well.

[5] *Location of New Component.*   If a mutant subpopulation is established bearing the new genetic material, what happens next depends largely upon where the subpopulation is located geographically within the overall parent population. Is it right at the geographical core of a large well-established population, or is it out at the geographical periphery in a more or less isolated subgroup? If

it is at the core of a dense population, then the sheer size and homogeneity of the group is likely to dilute the new genetic material in relatively few generations of breeding, and the mutant strain is likely to be re-absorbed.

If the entire parent population is relatively small or marginally established, with little diversity, a mutant strain that does develop at the core with even a slight physical advantage over the parent stock may displace the entire parent stock over a period of many generations. This is the sort of change that Darwin believed happened most frequently in the evolution of new species. It is called *phyletic* change.

The contemporary view, somewhat different from Darwin's, is that new species most often branch off from the parent stock at the geographical fringes, where the homogenizing influences are less strong. At the fringes there is diversity, which makes it possible for colonies of genetically altered individuals to develop. Over generations of further mutation in such colonies, new species are likely to branch off from the parent stock. This sort of change is called *speciation.*

The modern disagreements with Darwin are not over the scheme of natural selection, but over how rapidly new species develop and whether they develop more often by phyletic change or by speciation.[7] Both kinds of change occur, on either account.

[6] *New Problematic Equilibrium.*   Now for the possible outcomes of Darwin's plot, and in some cases the setting for further mutation.

First, the mutant strain may be displaced by its competitors at the geographical fringes of the parent population. The competitors may simply have characteristics that are more suited to the facts of life in the region than those of the strain we have been tracing, or they may be more flexible in one way or another. In difficult times, flexibility is important. A given number of bad seasons can select for an established strain by simply wiping out the mutant strain that is less able to cope.

Second, the mutant strain may persist at the fringes of the parent population, either by displacing its competitors in an exact reversal of the first outcome or, more likely, by coexisting for at least a time and increasing the total diversity at the geographical fringe. After many repetitions of the plot just outlined, speciation may occur.

Third, the mutant strain emerging at the geographical core without a distinct physical advantage is likely to be re-absorbed.

Fourth, the mutant strain may persist at the core, if it does have a distinct advantage that makes it better adapted than the parent population. This is the situation where phyletic change is likely to occur.

These four outcomes are too neat to be true, of course. It isn't often easy to decide what counts as the geographical core of a natural population as distinct from its periphery, or what counts as a greater or lesser degree of adaptation. That is one of the reasons why it is so difficult to fill in all the details of Darwin's plot with a fleshed-out story. Evolution takes place in complex animals at a very slow pace, so slow that the changes are barely noticeable as they are happening. When we become aware that a change has happened, the evidence as to why it happened is often long gone.

The organic operation of natural selection on human beings may seem to have come to a halt, but this is precisely because of the slow rate of organic change. Adaptive change can still be observed in human beings in the favoring of specific blood types and hemoglobin characteristics in given geographical areas. Such change is attributed to the continuing pressure of natural selection, in that the people in a given region who have the typical regional blood characteristics are also those who are most resistant to the diseases that are common in the area. Those with blood characteristics less resistant to the local diseases tend to die off at an earlier age.[8]

Any schematization of this sort has to be deceptively simple. It has to leave out the fact that mutations don't occur one at a time, and that in a given region all of the biota—animal and vegetable—are undergoing constant mutation simultaneously, and at varying rates of speed. If a given mutant strain is displaced at any of the exit points along the left-hand side of the schema in Figure 2-2, the situation doesn't just return to exactly what it was before this particular mutation occurred. The equilibrium is changed no matter what happens to any one mutation out of the many that can occur in a single generation of a single species.

The adaptation of a lineage of populations over many generations is not something that happens just to that lineage. It is best understood as an ecological *relationship* between the lineage—with its mutations that occur in a more or less random way—and everything else in the environment: the climate, the terrain, and the other life-forms of the region, which include potential competitors, potential predators, and a potential food supply.[9]

There is a cluster of problems, all having to do with survival, and indefinitely many mutations that can be viewed as trial solutions to those problems, some of them successful, but most not. It seems a gloomy prospect. A biologist will quickly point out that the usual outcome of mutational change is extinction.

In general, the individuals that do survive and breed are those which are better adapted to the local facts of life than those that don't survive. That is the simple key to Darwin's overall plot. But any successful adaptation is only one of many possible successful adaptations. There are many workable solutions to the problems of survival, as the enormous variety among living beings attests.

Now that scientists are able to study the genetic chemicals themselves, they find that even within species, all natural populations are extremely diverse. The tendency to diversity itself is favored by natural selection. It amounts to versatility in a collective, rather than an individual, sense. If a population contains within it a large number of variations, it is more likely, as a *population*, to be able to adapt to changes in its environment. However much the environment may change, in a diverse population there is a great likelihood that there will be at least *some* individuals which are well-adapted to a new situation.[10]

It is clear, then, that there are any number of different successful organic solutions to the basic problems of existence. There is no need to suppose that there is a uniquely best way for a population of living beings to deal with a set of problems, or that one very successful solution necessarily eliminates all others.

But each solution—each adaptive change in the characteristics of a population—alters the ecological relationship between the population and the total environment in at least a minimal way and thereby alters the character of the problems. A solution to one problem is thus likely to generate new problems.

Down to Cases: How Do People Get Information?

The plot I have just described is the plot for billions of separate but interlocking stories. Each story is about a single mutant individual, the result of a single genetic mutation, which can only result in a very minor difference from one generation to the next. A single mutation doesn't change things very much. It doesn't, by itself, ever result in the immediate replacement of a parent population or the emergence of a new species at the edge of a parent population.

By the best evidence, the history of life on this planet goes back for at least 3 billion years. It has taken many billions of small changes from one generation to the next for complicated organic individuals like human beings to develop.

Now, what do these billions of stories based on Darwin's basic plot have to do with the activity of doing science? They provide the starting point for an explanation of how and why people do science the way they do. For a start, the selection of physical characteristics explains ultimately how we come to have the characteristics we do, and these in turn make it possible for us to do certain things and impossible for us to do others.

The question isn't whether or not the sensory and neural characteristics associated with human thought and language are constrained by our genetic heritage but, rather, to what extent they are constrained. If we were studying the spinning activity of spiders instead of the theorizing of human beings, we would have to pay some attention to the physical equipment—the glandular mechanisms—that make it possible for a spider to produce a web of one sort rather than another and to the means that the spider has for consuming the food caught in its web. Both of these place constraints on the forms that an effective web can take. They steer the spider's activity on one course rather than another.

But we are talking about human beings, not spiders, and we are interested in that particular set of activities that constitute doing science. What I want to pay attention to now, therefore, is the sensory apparatus that ultimately provides us with something to do science *about*. How do we get the information that raises the questions that we try to answer scientifically? Why do we get one sort of information rather than another?

In the long evolutionary process, populations of living beings have developed sensory apparatus enabling them to intercept raw stimuli from the environment, as well as nervous systems and brains enabling them to interpret such stimuli in at least rudimentary ways and modify their behavior accordingly. Just moving to avoid danger, or to grasp at a bit of nourishment, requires their having a mechanism for detecting danger and nourishment, for differentiating between the two, and for moving away from the one and toward the other. The particular sort of sensory equipment that will be selected within a given population, after many stories on the Darwinian plot, depends upon where the information is to be had—on the sort of raw data available that will be of any use.

The most important sense for human beings is the visual sense. Far and away the larger portion of our brain is devoted to interpreting visual data.[11] The portions of the brain associated with the other senses are relatively small. In other species, the chemical senses (taste and smell) are more important. In still others, the means for detecting vibrations in the atmosphere (hearing) is more important than either the visual or the chemical sense. In some cases, the vibrations are originated by the organism itself, and the echoes are detected. This is true of both bats and aquatic mammals such as dolphins. Why?

Trace back through millions of stories based on Darwin's plot to answer that question. Where is the story set in a given instance? In the oceans, on dry land, on swampy land, on an open plain, in a dense jungle? How big is the organism in question compared to the local animal and plant life? What sort of nourishment can it make use of? What sort of dangers must it avoid in order to survive? Each of these questions, asked and answered over and over, fills out the story line for each minute change in a population of living beings.[12]

The human visual system is sensitive to a narrow portion of the spectrum of electromagnetic radiation, the portion we call, for obvious reasons, visible light. (See Figure 2–3.) The electromagnetic radiation most available on this planet from natural sources is determined by the characteristics of the sun, the distance of the earth from the sun, and the characteristics of the earth's atmosphere. It is the radiation in the range between 350 nanometers and 750 nanometers in wavelength (Figure 2–3), right in the area identified as the visual spectrum.

Radiation in the same area of the spectrum as visible light is generally useful in maintaining life-forms through the photosynthesis of plants and other processes. Radiation outside this area of the electromagnetic spectrum is generally harmful to terrestrial life-forms if it is induced artificially at the same intensity at which light occurs naturally.

Most organisms on earth have sensory apparatus which detects variation in direct and reflected electromagnetic radiation within the range of visible light. For human beings and many other species, it is the most important part of the sensory system.

There are many different ways in which organisms can detect light, ranging from "eyespots" in simple forms of marine life, which react simply to the presence of light, to the human eye and those of other highly complicated species. There is no luxury in nature, or hardly any. In general, no organism develops more visual and neural equipment than it needs to survive. The visual equipment of some species is quite primitive compared with ours. Mollusks, for example, have eyes that resemble the simple pinhole camera. Some insects have compound eyes which transmit many impulses through distinct tunnels, eventually to form an image.[13]

The human eye is a lens system—a *lenticular* eye. Such eyes have evolved independently at least four different times, in four different genetic lineages. Most vertebrates (including human beings) have

Figure 2-3 • The Electromagnetic Spectrum

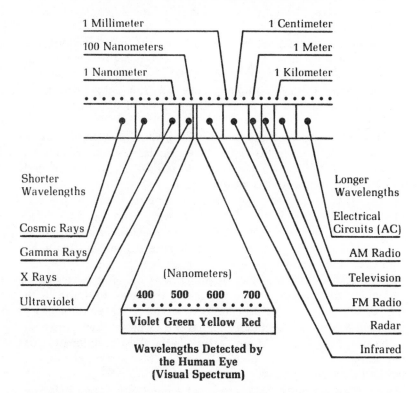

The increments represent wavelengths of electromagnetic radiation arranged on a logarithmic scale; that is, as you move from left to right, you multiply by 10 at each increment. The unit for measuring wavelengths of light is the nanometer, or millimicron, which is 10^{-9} meters, or 1 billionth of a meter, or about 1/25,000,000th of an inch. The visual spectrum is shown on the expanded diagram in linear scale.

lenticular eyes. But so does the octopus, and so do other species, which are not related closely enough to man to have a relevant common ancestry.

We tend to confuse ourselves about the visual sense in the way we talk about it. We sometimes think of our eyes as windows on the world. But who is looking out the window? It is easy to slip into misleading ways of putting things, such as "I see through my eyes," as if the "I" and the eye were quite distinct—as if "I" were somewhere inside my head peering out at the world through the organs at the front of my head with, presumably, *another* set of eyes. But the eye does not pass along every stimulus it receives. It selects. It filters. The difficult question is, Where does *interpretation* of visual stimuli begin?

Philosophers and scientists sometimes like to think of the interpretation of sensory data as something that is a higher function of human intelligence operating on "raw data" supplied by the sensory systems. What I hope to make clear is that "raw data" stops being raw at the surface of the eye and the exterior of the other sensory organs.

For the sake of this discussion, I will speak of *interpretation* of a stimulus as that process by which a particular portion of the total stimuli from the environment is selected by an individual and reacted to in one way or another.

Now that is a definition of *interpretation* from a psychophysiological point of view. It is unsatisfying from a philosophical point of view, because it doesn't make any mention of just where in the visual process consciousness, intelligence or concepts come in. Interpretation, we want to say, is something that people do, not something that nervous systems and brains do—just as seeing is something that people do, not something that eyes do. But as with the "I" and the eye, the distinction between a person and the sensory and neural apparatus that a person has is difficult to draw. A person is the whole package of items that we in the West traditionally sort out into "mental" and "physical" items. The sorting out hasn't worked very well, as I will argue later.[14]

I am not suggesting that there is not a line to be drawn between the unconscious, relatively physiological, filtering of sensory inputs and the interpretation of those inputs as information that enters into human conceptual and linguistic activity. The difficulty is that there are too many different places where the line can be drawn defensibly, any one of which would serve my purpose here. The line is at best a fuzzy one wherever it is drawn, even if we convert the question to one of consciousness. How much of the filtering and selection of visual stimuli can we control consciously? Even *that* is open to question, as are questions about how much we are able to control our physiological states consciously.

I am going to stay with the neurophysiological definition of interpretation here, because it serves the purpose of getting at answers to the two questions just raised: How do we get information through the senses and neural system? And, Why do we get one sort of information rather than another?

The filtering and organization of visual stimuli—the interpretation, as understood here—begins at the surface of the cornea (the outer surface of the eye) and continues through the nervous system and the brain. The lens of the human eye focuses an image on the retina, at the back of the eye. The image is upside down, like the image focused by a camera lens on a piece of film. Right away, some people want to ask where the image is turned right-side up, so that we can look at it properly. But that, again, is a confusion that grows from trying to distinguish between the "I" and the eye. We don't look *at* the retina as if it were a movie screen. The retina is part of us. We don't look *through* the retina at the world. We look at the world *with* the retina and the rest of the visual system.

The retina is an encoding device. It does not supply neutral stimuli to be encoded by something else. It interprets light physically by converting it into bioelectrical energy, which is the pulse-coding of the nervous system. The determination of what portion of raw visual data is transmitted beyond a given point in the nervous system, and how it is organized for transmission, takes place on a distinctly physiological level in the nervous system itself. In the retina, for example, there are 126 million receptors of visual stimuli (about 6 million cones and 120 million rods) but only 1 million channels (ganglion cell axons) for transmitting the encoded stimuli beyond the eye itself. There are several other points in the neural system where such filtering and organization takes place on what can be identified as a purely physiological level.[15]

But the physiological data about how human beings obtain and interpret information begins to thin out at this point. We know a good deal more about how visual data is filtered and organized in the nervous system than is really relevant here, but we know less than most people think we do about just how that data is processed in the human brain.

We have to work at this last physiological question from a different direction. There is an indirect way of getting at the characteristics that all human beings have in common. It involves an inferential step that moves us from one discipline to another.

Let me explain, in general terms, what happens when we run out of one kind of evidence about a species and have to rely on another. The question is, what characteristics of a population are due to its genetic heritage? Most of the answers to this question come from physiological studies. Where the physiological information is not clear, we look for the answer in the abilities of members of the species to do certain things or in the limits of such abilities. We arrive at the answer in several distinct steps.

If all members of a given species of bat, for example, are able to guide their flight by means of their sonic echoing sense—they emit high-pitched squeaks and gauge the distance and position of objects by the echo—we say that the ability is *species-specific*. To tell which traits are species-specific, we have to sort out the traits of individuals which are learned or acquired from those which are common to all members of a species, whether or not they can acquire the trait in a natural setting. This entails isolating and testing individuals where possible and careful observation of large samples of a population where isolation isn't possible.

Now here is the inferential step that moves our questions from physiology to experimental psychology: If a given trait, ability, or constraint is species-specific, we infer that it is physiologically based and therefore part of the genetic heritage of the species, *whether or not* we are able at a given time to describe the physiological processes associated with the trait.[16] Human abilities to discriminate color (hue) provide a good example here and lead us back to the mainstream of questions about change.

A human being whose vision is statistically normal can distinguish as "different" (in hue) two patches of color that reflect wavelengths of light which differ by as little as one to 6 nanometers. (See Figure 2-3.) The range of radiation that we can detect extends at least from 400 nanometers to 700 nanometers in wavelength. That means that we can distinguish, under good circumstances for comparison, more than 200 different hues. This is far more than any human language has names for.

That isn't all we know about human color discrimination, of course. We know some of the physiology as well. The signals transmitted from the retina are determined by the interaction of given wavelengths of radiation with "pigments" in the rods and cones at the back of the retina. Here the "color signal" is encoded, and here is where we become unsure about the remainder of the physiological processes involved.[17] We have to look for further information at the other end of the process by finding out just what sort of discriminations people can make and report.

Individuals differ somewhat in the color discriminations they can make, which is why I had to refer to statistically normal vision above. Beyond this, it is likely that there are individual differences in the pigmentation of the rods and cones of the retina, so that subjective color experiences vary greatly from one individual to another. There is really no way to tell about this matter. Some people think it doesn't even make sense to *ask* whether or not our subjective experiences differ when we look at patches that reflect exactly the same portions of the visual spectrum. Whether or not you and I see the "same color" when we both look at the blue sky is something that we can't tell. We *can* tell whether or not we both agree that the sky is blue. What is important in determing whether or not a person has normal color vision is whether or not he can learn to react to specific visual stimuli by naming the colors in appropriate ways. The names for colors are typically among the first words that we learn as infants, and the particular colors that we learn to name are matters of language and culture.

What is true of vision in this respect is true of the other senses as well. We *can* make more discriminations than we ever do. Physiologically, human beings are able to discriminate more variations in shade or hue than any human language has names for. Some languages have as few as two or three names for colors.[18] But clearly this doesn't mean that people who speak those languages can talk about only two or three hues. There are always modifiers to increase the color distinctions that we can talk about. We easily and clearly speak about red, orange, and reddish orange in English, for example, filling in further distinctions between the major distinctions that we make. In any case, a person who speaks one language is capable of learning another which includes distinctions that his native language doesn't, and of making distinctions properly.

Language is an encoding device. So is the retina. Both encode certain bits of "information" and not others. The retina encodes visual stimuli to be interpreted further along the line of the psychophysiological process. A language allows us to encode information to be interpreted by other people. But we can't let the analogy run away with us, even though we might want to say that we "inherit" both our retinas and our languages. The characteristics of the retina place limits on what we can see—we can't see outside the visual spectrum—and they also make it physiologically possible for us to make discriminations among the stimuli that come to us. But we cannot change the characteristics of our retinas.

Similarly, the characteristics of the language that we speak constrain the kind of information we can pass along to other people while at the same time making it easy to pass along certain kinds of information. Within what the physical equipment makes possible, a given language *selects* those features of our experience that we name and talk about. But we can always learn a new language, and this is an important difference.

Using language is something we learn to do. It is an activity, and there are many habits that have to be acquired in learning the activity. Some are conceptual habits, which have to do with recognizing and distinguishing features in our sensory experience. (Here we are in the problematic territory where philosophers and psychologists try to draw that difficult line I mentioned earlier.) Others are linguistic habits that must be learned in order to communicate with each other about our experience as we conceptualize it.

Habits are tendencies to behave in specific ways. Linguistic habits in particular are passed along from generation to generation. They change from time to time in a given population. They differ in marked ways from one population to another. Does this begin to sound familiar? We are talking about change again—change in habits of behavior. It is time to take another look at Darwin's basic plot.

The Natural Selection of Behavior Patterns

Each of the stories based on Darwin's central plot for bodily change has a sequel, at least in the animal kingdom. Animals don't just lie there and vegetate. They *do* things. And different populations of animals do different things, even within a species.

There are two main phases of evolution in organisms—

endosomatic evolution, which is the modification over a period of time in the bodily characteristics of a population, and *exosomatic* evolution, which is the modification over a period of time in a population's tendencies to behave in specific ways. *

As with any distinction of this sort, there are clear-cut cases on each side and a somewhat fuzzy area in the middle. On the endosomatic side, clearly, are the development (over many repetitions of Darwin's plot) of such things as dorsal fins, skeletal systems, opposable thumbs, nervous systems, and sensory organs. On the exosomatic side, clearly, are spiders' web spinning, beavers' dam building, and birds' migrating. The fuzzy area includes, for the most part, those habits of behavior which have to do with an organism's use of its own body: techniques for locomotion, or for focusing vision, or for "using" the nervous system in given ways. In such borderline cases, we can take it as sufficient evidence that a trait is exosomatic if two individuals which do not differ significantly in physical characteristics differ with respect to the behavioral trait. Such traits are clearly not species-specific. Different local populations of many species vary in their food preferences and hunting techniques, and in the higher mammals they differ in the ways they organize their bands or colonies.[20]

We have to distinguish among three things:

The ability to behave in a given way,
The tendency to behave in a given way, and
The actual habits of behavior.

The ability is physiological: endosomatic. You can't do what you lack the physical equipment for. Or, to put it more positively, an organism can do whatever it possesses the bodily preconditions for. We *can* distinguish more than 200 hues within the range of our vision. We *can't* see X rays. This is an organic matter. The spider can spin its web only because it has an organ called a *spinnaret*.

The tendency to behave in specific ways lies in the fuzzy area between exosomatic and endosomatic characteristics. I will have more to say about this as we go along. The actual habits of behavior that an individual develops are clearly exosomatic. Even if a given behavioral tendency is inherited, it can be defeated by circumstances, and the behavior may never take place.

---

*The term "exosomatic" is used most prominently by Sir Karl Popper to identify what biologists or physical anthropologists might also call nonorganic or superorganic biological adaptation as developed by a particular evolving lineage.[19]

The behavioral variations possible within a given set of physical constraints—the exosomatically evolved characteristics of a population—tend to change faster than the bodily (endosomatic) traits. We can actually trace the history of a good bit of the exosomatic development of human beings and other species.

Behavioristic psychology, particularly as represented by B. F. Skinner, involves the study of the human species as a biological system. A behavioral psychologist speaks quite freely of the evolution of behavior patterns. To get at how and why populations of human beings develop characteristic patterns of linguistic and other behavior, I want now to look at how Skinner in particular elaborates Darwin's basic plot.*

[1] *Problematic Equilibrium: Established Repertoire of Behavior.* Here we must narrow the field of focus again to individuals. At any given time in its life, an individual has a repertoire of established behavior patterns. They might be quite primitive—breathing, swallowing, seeking nourishment in simple ways—in the case of infants or of simple organisms. But human beings and other highly developed species have more complex repertoires of behavior that include ways of breeding, raising the young, participating in cooperative group behavior in defense, seeking food and shelter, and so on.

Behavior patterns, like bodily characteristics, develop out of the total relationship between the individual or population and the environment. They also constitute a set of solutions to problems of survival, at least in part. The established repertoire of such patterns constitutes the background into which any new behavior must be integrated.

[2] *Innovation: Experimental Behavior.* There seems to be a genetic basis for the tendency to experiment with innovative behavior and to retain or abandon it according to the outcome.[21] Individu-

---

*It has been argued by many people that behavioral psychologists, particularly as represented by Skinner, fail to give an adequate account of human emotions and intellectual activity, precisely because they tend to view complex verbal and intellectual behavior in terms of relatively simpler nonverbal behavior. But the question I am after here is how behavior patterns develop at all, and Skinner gives an excellent account of this, especially in *About Behaviorism*, where he takes into account the criticisms of his overall thesis that accumulated over a period of twenty-five years and gives a clear and straightforward account of the behaviorist approach as it stands now.

Figure 2-4 • The Natural Selection of Behavior Patterns

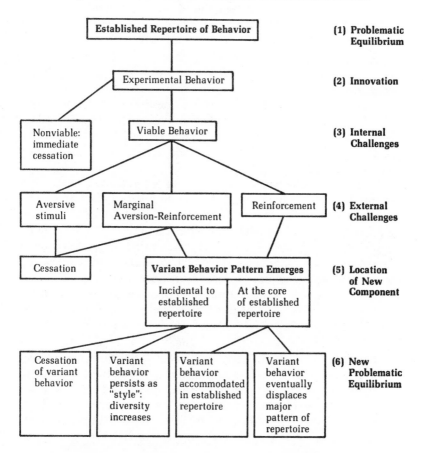

als or populations that lack such a tendency literally cannot learn and cannot adapt their behavior to changing situations; they die out as circumstances change.[22]

Behavioral innovations happen largely as a matter of chance, as do the genetic mutations discussed earlier. Skinner holds, essentially, that experimental behavior happens in a more or less random way *before* any function can be attributed to behavior. The appeal to Darwin's overall plot is an attempt to explain why some random experimental behaviors are preserved.

[3] *Internal Challenges: Contingencies of Survival.* If experimental behavior does occur in random ways, then some of the experiments

will be abandoned immediately because they are disastrous. The result is the "extinction" or cessation of the experimental behavior, perhaps accompanied by the death of the experimenter.

If the particular experiment proves viable from the point of view of the organism itself, the possibility arises that the experiment will be repeated and that the experimental behavior might eventually become part of the repertoire of an individual and perhaps of an entire population. Now here is a crucial question: How can we move from the experimental behavior of an individual to the consideration of whether or not the behavior might be adopted by a whole population? We do not have a genetic mechanism here to pass characteristics along.

The answer lies in the notion of *selective pressure*. The facts of life in an individual's environment determine which behavior will be disastrous and which will not. The same facts of life exert pressure on every member of a population, although perhaps not to the same degree on every individual. There is maximum pressure on a population when there is a sudden change in the environment and the established techniques for sustaining life are no longer effective. It is here that species-specific traits undergo change or replacement.

Time for an example. Imagine a species of mammal that eats only one kind of plant. The plant is plentiful and nourishing, but there are other plants available which are equally so. A blight, fungus, or other natural agency wipes out the plant that has been the population's sole food supply. Individuals that possess the tendency to innovate and try other plants as food will stand a chance for survival. Individuals that do not possess the tendency will not. They will fail the first test as soon as the established food supply is wiped out. Their lack of a genetic tendency to innovate will cause their extinction.*

Some of those individuals that *do* innovate may be wiped out immediately because the new foods they try will be poisonous. Others will find new foods which are both safe and nourishing.

The pressure in the example is extreme, but it is not at all implausible that such a situation has happened many times in the history of animal life. The details of the story—the disaster that

---

*If this seems farfetched, remember the koala bear, a marsupial that will eat only one variety of eucalyptus leaves. They will starve to death if those particular eucalyptus leaves are not available, even though other plants—including other varieties of eucalyptus—are plentiful.

destroys the established food supply, the fact that some of the plants tried as alternative foods might be poisonous—are what Skinner calls *contingencies of survival.* Contingencies of survival are the facts of life—things which happen to be the case concerning the individual or population itself and, more important, its relation to the total environment, including all the surrounding biota.

Contingencies of survival produce species-specific behavior, which is also called *innate* behavior or *instinct,* common to an entire species. Now this is a tricky point and worth dwelling on for a moment. I said earlier that species-specific traits are genetically based, that they are typically physiological constraints on behavior, and that they may be abilities or tendencies to behave in specific ways that are rooted in the nervous system. But behavioral traits *become* species-specific; they don't start that way. Physiologically based behavioral traits are *selected* from among the diverse genetic traits that arise within a species by genetic mutation. In the case of our hypothetical mammals whose food supply was changed, the contingencies of survival would have established a tendency to experiment with new foods as species-specific simply by wiping out all those members of a population which did not possess the tendency.*

Not all behavior traits are the result of the strong pressure of survival, however. In species that have a strong tendency to innovate and possess the luxury of not living at the edge of survival from a purely physiological point of view, further selection of complicated behavior patterns goes on in ways that are not linked causally to the genetic heritage of a population. The possibility develops that there will be great diversity in habitual behavior within a species, and even within a population within a species. The distinctions I make within the human species will be drawn on the basis of behavioral variations that do not ordinarily put the survival of whole populations at stake.

[4] *External Challenges: Reinforcement and Aversion.*   Whether or not experimental behavior is repeated, and whether or not it

---

*A recent and controversial development in the scientific literature surrounds Edward Osborne Wilson's book *Sociobiology.* The controversy is over this question: Just how much of human social behavior is genetically constrained? Wilson holds that much of it is, down to rather minute details. His critics hold otherwise.[23] The matter is far from settled, but I think the evidence is for Wilson's critics. Beyond the point where survival is at stake, the genetic tendency is always toward diversity.[24]

eventually becomes part of the established repertoire of an individual and perhaps a population, depends upon the consequences of the behavior. These consequences in turn depend upon the circumstances: the facts of life in the total environment plus the previously established behavioral repertoire.

Some behavior has positive consequences and some does not. When a hungry animal behaves in a way that gets it food, for example, the behavior is *reinforced* by that consequence and is therefore more likely to recur.[25] Negative consequences are called *aversive*.

In large measure, aversive stimuli are closely related to survival issues. Think of pain, which is the most obvious case of an aversive stimulus. It is a matter of survival that most species have a means of detecting danger before it becomes a literal threat to life. The neural system that generates pain as a warning of too much heat or cold, of breaks in the skin, or of irritating chemical agencies, is part of the survival mechanism. But specific reactions to such pain signals might not relate so closely to actual threats of death as does the physiological mechanism. If the organism's behavior is flexible enough, aversive stimuli may well cause the cessation of the experimental behavior without causing the death of the experimenter.

The behaviorist thesis is that in the event of reinforcement the random innovative behavior is likely to be repeated and, in the event of no reinforcement or diminished overall reinforcement as compared with what it was before the experimental behavior (for example, if the experiment results in getting less food than other behavior does), the innovative behavior is "extinguished" and recurs rarely, if at all. And, if it does recur, it recurs by chance.[26]

Conditions that provide reinforcing or aversive stimuli are called *contingencies of reinforcement*, and they are exactly what the term suggests: things which happen to be the case—the facts of life again. But this time the issue isn't the survival of the organism, but the preservation of the innovative behavior. The facts of life exert a selective pressure that preserves advantageous behavior by reinforcing it and tends to extinguish disadvantageous behavior by not reinforcing it or by generating an adverse result.

[5] *Location of New Component: A Variant Behavior Pattern Emerges.* Whether or not a variant pattern of behavior is established at all depends upon the contingencies of reinforcement.

How an established pattern affects the overall repertoire of an individual or a population depends upon how central or how incidental the new behavior is to the total repertoire. How much the overall repertoire of established behavior—the behavioral equilibrium between the population and its environment—is affected depends, in other words, upon how closely related the new behavior pattern is to the central issues of survival.

[6] *New Problematic Equilibrium.*   Again, there are four possible outcomes that we can distinguish. The dietary innovation in the population of mammals discussed at step 3 (internal challenges) was mitigated by a sharply defined issue of survival. In Figure 2–4, this behavioral change can be traced down the right-hand side of the schema of selection. The dietary change would have maximum survival value (3); it would, by the sheer force of this, be strongly reinforced (4); it would be relevant to the central core of the population's established behavior (5); and it would displace a major feature of the established repertoire. The ecological equilibrium between the population and the surrounding biota is altered by the fact that this particular population now eats different plants than it used to.

Other innovative behavior, although it may be closely related to the "core" of the species' repertoire, can be accommodated into the overall repertoire without displacing major elements. The kind of nourishment an animal needs is a physiological matter; the selective pressure will be very strong to establish a species-specific tendency to favor certain kinds of foods over others. But the particular technique for obtaining that food may vary considerably within a species or even within a population.[27] There may be a number of efficient hunting techniques, for example; they may or may not be equally efficient in different terrains.

Still other behavioral variations may persist as "styles" of incidental behavior within a population. For the most part, such stylistic differences from one individual or subpopulation to another occur in that portion of the repertoire that has the least direct importance to survival, where neither reinforcement nor aversive stimuli are particularly strong, and where they are not exerted in the same way on an entire population.

In natural selection, whether of bodily features or of habitual patterns of behavior, the appearance of innovations is a crucial part of the plot. Once they appear, innovations are selected by the pressure of the prevailing facts of life.

Down to Cases Again: How Did Language Develop?

Bear in mind that the schema of Darwin's plot is very simplified, whether we are considering Darwin's elaboration of the bare bones schema to explain bodily change or Skinner's elaboration of it to explain behavioral change. Evolutionary change takes place within a species in a much more complicated setting than any schema can take into account. Moreover, when we are interested in very complex patterns of behavior like human use of language, we must talk about billions of individual stories based on the same overall plot, taking place over hundreds of thousands of years.

Human beings, like many other species, have been under great selective pressure to organize into social groups. For such "social animals," the group serves as a sanctuary and a stronghold, as well as a means of meeting the needs of individuals by a division of labor into specific tasks that some individuals perform on behalf of the entire group. There are other familiar examples of this among the higher mammals as well as among fish and insects.

By the usual estimates, human languages developed about 40,000 years ago. The kind of selective pressure that led to the development of languages in human beings has also led to complex vocal (not verbal) behavior in other land-based species. Among all social animals, individuals mimic each other's cries, and there are familiar cases where a species has developed distinct cries which carry specific meaning: "danger," "food available," "all is well," "get out of my territory," and other such basic bits of information or exhortation.[28]

Any behavioral development which makes the transfer of information by vocal or facial expression more explicit is subject to great selective pressure. It increases the advantages of living in a group by increasing the efficiency of the group in dealing with its environment.

From an anthropologist's point of view, the sequence of developments that led ultimately to symbolic systems and language can be summarized as follows: attempts by individuals within a group to influence the behavior of other individuals led to the association of specific reactions with sounds. This already involves many stories that follow the basic behavioral plot. Further, the pressure of natural selection to increase the efficiency of the group led to the sophistication of vocal imitation and of vocal control that is common to all human beings. Many more stories, same basic plot. The capacity to learn language is a species-specific trait of human beings.[29]

How much of what we call *consciousness* entered into the development of languages at the early stages is impossible to say. But this much is clear: it is the reinforcing pressures of living in communities that led to the development of grammatical patterns of verbal behavior. The established habits of a population are among the facts of life for individuals born into the population. Reinforcement happens within the community, in the teaching of the young, as well as in the relationship between members of the community and the overall environment. The grammar and natural syntax of the linguistic behavior of a population are things that we discover after the behavior has developed. They are "extracted" as rules from naturally evolved activities in much the same way that a naturalist might set down a system of rules for a beaver's dam building in a systematic way after studying the activity closely in its natural setting.[30]

The Differences Among Language Communities

There are obvious characteristics that vary within a species over periods of time, over wide geographical areas in which the species lives, and over the different sorts of terrain and overall ecology that the species inhabits. These variations are of little or no use to the biological taxonomist, who seeks to classify groups of organisms on purely genetic grounds. The taxonomist is interested in those features which are common to whole species or subspecies.

But evolutionary change doesn't stop with the development of a species. Adaptive change goes on within subgroups within a species, often in a way that does not permit the isolation of genetically linked populations, precisely because of the mobility of individuals within the species and the fact that, as the term "species" is defined, members of the same species can interbreed, whether or not they are from the same population.

The general term for the kind of variation I want to pay attention to now is *cline*.[31] The easiest illustrations of the use of the term have to do with bodily characteristics, although I will want to apply it to behavioral variations as well. There is a bird called the coconut lory, native to New Guinea, that provides an example. Individual birds are slightly larger in size as one moves from west to east through the range of the species; then, at the eastern extreme of the species' range, the birds are suddenly smaller again. Also, as you move from north to south through the species' range, the blue color of individual birds is brighter.[32]

These are clinal differences, and from the example it becomes clear that a given individual may be taken to belong to a number of different clines, depending upon which characteristics we pay attention to—in this case, depending upon the angle at which we travel across the bird's range. Clinal differences are just as evolutionary as the differences that produce neat taxonomic classes. They have to do with adaptive change within a species. A given population, and a given individual within a population, can belong to only one subspecies, and a given subspecies can belong to only one species, and so on. These are matters of genetic classification, taxonomy. But an individual or a population of the same species may belong to a number of different clines.

Many species, including human beings, have wide geographical ranges. Geographically separated populations within such species differ from one part of the range to another in the specific details of the relationship that individuals develop to the environment. Surely the core populations of the several human races can be viewed as clines. They are populations which differ from each other genetically at the core, typically because of clearly identifiable geographical aspects of their history. As human beings spread out over the entire globe, populations became localized—relatively isolated from each other by mountain ranges, as between India and the Far East, or by settling on different continents. But the genetic lines among races are not, and never have been, lines that can be cleanly drawn. They are gradations, like the characteristics of the coconut lory, because members of the species move about, interbreed, and establish new colonies. Behavioral characteristics of populations can be usefully viewed in the same way.

Distinct linguistic styles developed in separated populations because the things that members of those populations needed to communicate with each other about differed from one location to another, depending upon the local conditions of life. Now once a given pattern of linguistic behavior is established, it becomes the setting for further development. We are back at the beginning of the plot for behavioral selection. The established repertoire of linguistic behavior in a given cline characterizes the problematic equilibrium at a given time. All further development of language in the population and its descendants is influenced by that repertoire, which provides the structure to be altered by innovation and selection. A line of descent develops in linguistic behavior that can be traced in much the same way that the descent of bodily change is traced by biologists.[33]

There are written remnants of extinct languages that supply us with evidence of the character of linguistic behavior at a given time and place. Such records help the linguist to fill the gaps between existing language clines in essentially the same way that fossil remains help the biologist to fill the gaps between existing species.

Languages differ in their grammatical patterns and vocabularies—in what it is easy to say and in what distinctions must be attended to in constructing a sentence (for example, whether an object is animate or inanimate or an action completed or in progress). At the same time, populations have changed and mixed within the few thousand years of deliberately recorded human history, and the characteristic linguistic behavior of the inhabitants of any given location has changed. The human species has a global range. When we attempt to distinguish the linguistic characteristics of distinct populations within the human species, the connections among linguistic clines become complicated and fascinating.*

Recently, scientific interest has developed in the study of linguistics and psycholinguistics.[34] There are relatively few linguistic clines that have not been influenced by the rapid development of transportation and communication within the past few centuries, and those that have remained isolated are typically in small, geographically remote, populations. But they provide data for empirical study. It is fair to say that one of the central aims of modern linguistics is to determine whether or not there are structural features that are common to all human languages and that are likely to reflect some basic features of human neurophysiology. Recall Skinner's belief that species-specific behavior patterns, at whatever level of analysis they are detected, are physiologically based.[35]

*Most dictionaries list the historical roots of modern Indo-European languages, illustrating many times over the way in which linguistic activity changes in a given region.

Facts about human sensory apparatus and abilities, and such facts as may emerge from empirical studies of human conceptual and linguistic abilities, are contingent facts about the world. They could have been otherwise if the circumstances of our history as a species had been different. But these are the facts that make such experience *human* experience. In that sense they are necessary components of human experience. This just *is* the sensory and neural equipment that the species has, and our physiological characteristics shape both the sensory discriminations that we can make and the ways in which we are able to interpret and organize our experiences conceptually.

The best available evidence for how people think is to be found in what they say—in their conceptual and linguistic behavior. It is reasonable to expect that we will be able to identify a base which is common to all human thought and which reflects the character of the sensory and neural equipment that all human beings have in common.

That there are such basic categories and concepts seems clear. Attempts to describe such species-specific characteristics generally include the apparently universal human notions that physical objects exist in three-dimensional space, that time is irreversible, and that every event has a cause. Whether or not a specific listing of such categories is adequate becomes a matter for critical evaluation.*

But if there are basic categories of thought common to the whole species, they do not provide us with a way of describing or explaining our experience that is common to the whole species, as I argued in Chapter 1. How, then, do we learn the "correct" ways to describe the world? There is no single correct way for all human beings. Each individual has to learn the correct way to communicate with those around him from the community in which he lives. Correctness, in this context, can be a very local matter.

It isn't, therefore, just minor physical variations and cultural habits of community organization that differentiate populations of human beings, but linguistic characteristics as well, and, perhaps more importantly for this discussion, "habits of thought."

Among different language and cultural groups, there are noticeable differences in the habits of thought that are reflected in ordinary conversation as well as in more formal organized systems of

---

*Perhaps the best known such attempt to describe the human conceptual base was that made by Immanuel Kant, utilizing what was known about human perception and reasoning in the eighteenth century. A more modern version of the Kantian view has been developed by P. F. Strawson.[36]

knowledge. These differences are as basic as the color distinctions already mentioned, but they extend more deeply into language than vocabulary. There are tenses in some languages that don't exist in others (the aorist in Greek, for example, has no English counterpart), and some languages seem to be totally "untensed." European languages tend to focus on *things:* statements about things and their properties provide the model for most other statements, so definitions, love affairs, headaches, and streaks of good luck are all treated grammatically in the same way as physical objects. Other languages would treat all those items as activities—it would be more natural to speak of the act of defining than of the definition.

The differences don't, it seems, go as far as the basic organizational notions of space and time or the closely related notion of an event. These, as I have noted, seem to be physiologically based. There may be a few mutants among us who can "picture" objects in four-dimensional space and imagine nonlinear or symmetrical time. Surely the endosomatic preconditions—the neural apparatus—for such picturing would have to be present, and beyond that, such a mutant would have to run counter to the current of his own culture in order to learn to use his unique equipment.

Once again, it is appropriate to close a chapter about human characteristics with some musings about what other intelligent life-forms might be like. We may wonder what it would be like to experience four-dimensional objects or reversible time, or we may wonder what sorts of problem we might have in communicating with intelligent beings who evolved under conditions where they had to organize their perceptual experience in four or more spatial dimensions. We, as a species, obviously had no need to do so; we did not develop the equipment to do so; we cannot choose to do so even if we want to.

Beyond this wondering, which is not altogether idle, we might consider the notion of *event* that is basic to our discourse—something happening in three-dimensional space at a given instant of time. What differences might there be between an intelligent dolphin's basic notion of *event* and our own? What sort of language would it take to describe the sorts of event experienced by an intelligent being who, in addition to the sensory differences noted earlier, is never stationary, who always moves around the objects it focuses its attention on? In the case of life-forms that might have evolved on other planets and have different sensory ranges, what sort of basic organizing concepts might we expect from beings whose primary sense reacts to, say, X rays or to what we call microwaves? What could the concept of "solid object" be?

In the next two chapters, I will focus more sharply on the linguistic means that we have for making sense about the world and the activity of developing scientific explanations for what we experience. Any further discussion of alternative forms of intelligence, whether human, aquatic, or extraterrestrial, must wait until we have looked more closely at the basis of western science.

FURTHER READINGS

On natural selection, David L. Hull gives a clear explanation of evolutionary theory and some of its problems in *Philosophy of Biological Science* (1974), Chap. 2. For the differences between contemporary evolutionary theory and Darwin's, see Stephen Jay Gould, *Ever Since Darwin* (1977), and, for a clear and straightforward account of the implications of the theory, George G. Simpson's *The Meaning of Evolution* (1951) is excellent.

On the human visual system, *Light and Vision* (1966), by Mueller et al., is clearly written and well illustrated. On a more technical level, but thoroughly readable, is G. Hugh Begbie, *Seeing and the Eye* (1973).

On behavior patterns from a psychological point of view, B. F. Skinner presents the behaviorist thesis as it stands after twenty-five years of critical debate in *About Behaviorism* (1974). From an anthropological point of view, see Dobzhansky's *Mankind Evolving* (1962), or Buettner-Janusch's *Origins of Man* (1966). On a less technical level, Bronowski gives a brief treatment in Chap. 1 of *The Ascent of Man* (1973).

On the connection between sensory stimuli and concepts, H. H. Price, in *Thinking and Experience* (1953), gives a philosophically sophisticated account. A good sampling of twentieth-century treatments of the topic is to be found in Robert J. Swartz (ed.), *Perceiving, Sensing, and Knowing* (1965). See, especially, the final essay by A. M. Quinton. A more current treatment is in Joseph Margolis, *Persons and Minds* (1977).

On the concept of a person, see Margolis (above) and P. F. Strawson, "Persons," reprinted as Chapter 3 of *Individuals* (1963), and *The Bounds of Sense* (1966).

On the development of language, Dan Slobin, *Psycholinguistics* (1971), gives interesting examples of differences among languages. The important and interesting work of Noam Chomsky has provoked considerable reaction in both psychology and philosophy. See his *Language and Mind* (1968), for a good sense of the undertaking.

On nonhuman language and communication: Konrad Lorenz, *King Solomon's Ring* (1952). On dolphins in particular, John C. Lilly, *The Mind of the Dolphin* (1969), esp. Chap. 5.

On human evolution in general, the sources mentioned for the evolution of behavior patterns are excellent. See also Colin Wilson, *The Occult* (1973). Wilson's criticisms of evolutionary theory are in his chapter titled "The Evolution of Man," which begins on p. 121. In this connection, see also the classic study by Robert Ardrey, *African Genesis, A Personal Investigation into the Animal Origins and Nature of Man* (1961).

On the evolution of human intelligence, see Carl Sagan, *The Dragons of Eden* (1977).

THREE
On Making Sense

Supposing, Believing, and Asserting

Science is essentially a public activity, and for that reason it is essentially a linguistic activity. Beliefs are shared, and passed along from one generation to the next, by means of speech and written language. Beyond this, language molds our beliefs in two ways. First, the particular language we speak and the particular vocabulary within that language that comes naturally to us provide habitual modes of sorting out and describing our experiences. And, second, the use of language makes it possible for us to discuss each other's beliefs critically and to evaluate our individual beliefs against the background of a critical consensus of beliefs.

Even before we come to communicating and criticizing beliefs, the use of language enables us to suppose—to form hypotheses that we neither believe nor disbelieve at the outset—and to try them out publicly. Critical discussion of what we suppose hones and refines what we come to believe.

Any exchange of information among human beings amounts to an assertion, or series of assertions, that something is or is not the case. You can, of course, make assertions without using language directly—by nodding your head, pointing or otherwise gesturing, flipping the lever on your car's directional signal, or pressing a button. But words, gestures, or mechanical signals cannot be taken by themselves unambiguously. Speaking or gesturing must happen within a context in which it will be understood if we are to communicate our beliefs.

Logical questions can be raised within a well-understood context, to evaluate specific lines of reasoning, or in more rigidly defined contexts, to spell out the logical consequences of a particular statement of the laws and postulates of a theory. Logic is not, in general, concerned with hypotheses, beliefs, and assertions, but with statements (or propositions or sentences) which are assumed to be straightforwardly true or false, and with the evaluation of individual patterns of inference or argument. Such patterns are abstracted from the activities of asserting and critically discussing our beliefs. They are patterns that we detect in the "products" of the activities— the statements or sentences—that I likened earlier to the webs that are the products of a spider's activity.

To evaluate a set of statements logically we must assume that they are all made on the same common ground—from the same point of view. Recall the example in Chapter 1 where we considered the variety of reasonable ways of counting and describing the objects in a given physical space. The objects on my desk at this moment can be counted in at least three different and defensible ways. The outcomes are seven objects, or twelve, or sixteen, depending upon what you choose to take as a single object. Logically, the three following statements are mutually exclusive:

There are precisely seven objects on my desk.
There are precisely twelve objects on my desk.
There are precisely sixteen objects on my desk.

No two of them could be true. But each of the three is accurate on its own ground—that is, within the context of a given way of individuating objects.

Each way of counting yields a different inventory of the objects on my desk (five keys and a ring on one inventory and one ring of keys on another, for example). Moreover, each of the three inventories is, on its own ground, an accurate inventory of the items on my desk. I know perfectly well that there are seven items on my desk. If I were to lay out for you how I counted them, so we were on common ground, you would get the same result. I also know perfectly well that there are other ways of sorting out the objects that would result in counting different numbers of objects.

Now this, in miniature, is the situation we face when we consider alternative descriptions of any state of affairs or alternative explanatory assertions about why a given state of affairs is the way we describe it. We make assertions in such a way as to relate our own experiences to a common ground of conceptions that other people can understand—a point of view that makes rational discourse possible. Within such a point of view we can carry out the logical abstraction of statements and inference patterns, and discuss the effect that the truth or falsity of one statement would have on the truth or falsity of another made on the same ground. But frequently we have to step back from a particular context of assertion and explain or negotiate that common ground in order to be understood, prior to raising logical questions within the context.

Presuppositions

If I expect you to understand my assertions, I have to follow rules for making sense. Some of these are structural, linguistic rules that we absorb as we learn how to speak and write. Others characterize conceptual common ground on which we carry out rational discourse. The latter are *presuppositions* which delimit the contexts in which we understand, compare, and debate each other's spoken or written assertions. Changing problems, changing conceptions, and changing interests continually redefine the ground on which we assert and discuss our beliefs. Such changes also continually redefine the territory of the sciences on which the specialists carry out the critical debate that leads to consensus in the scientific community.

The presuppositions which characterize the context of common sense discussion change over a period of time and vary somewhat from place to place. Within what may be called the overall context of common sense, we can shift from one set of presuppositions to another, depending upon the occasion, or whom we are addressing, or the particular purpose of our assertions. This is exactly the sort of thing that happened in the several different enumerations of the items on my desk.

Whether we are talking about the loose and informal contexts of common sense or about the more rigorous contexts of scientific theories, we can distinguish several different kinds of presupposition which delimit the common ground on which we make and understand a given set of assertions.

Formal presuppositions are those I identified in Chapter 2 as descriptive of species-specific facts of human experience. Among them are the following:

All objects exist in three-dimensional space.
No object both has and fails to have a given property.
Individual objects occupy distinct, bounded, uninterrupted seg-
  ments of space and time.
Time is irreversible.
Every event has a cause.

Such presuppositions constitute the foundation of common sense
experience. They have remained unquestioned in a very solid sense,
although, as I noted in Chapter 1, there have been some interesting
suspensions of them for purposes other than describing experience:
contemporary physicists view gravity as a curvature of three-
dimensional space in a fourth dimension, speak of reversible time
and causal sequences at the subatomic level, and allow that subatomic particles might violate the general presuppositions about
individual objects. In general, we do not suspend formal presuppositions except at the far reaches of theoretical explanations. The
exception is, of course, to be found in certain types of science fiction
or fantasy, where the experiences of characters might include
multidimensional space and travel "through time."*

Ontological presuppositions concern the kinds of thing we are
willing to claim exist. This broad group of presuppositions can be
broken down into subgroups in several ways. For now, we can
observe that ontological presuppositions may be of a very general
sort such as the following:

Physical objects exist.
Minds exist as distinct from bodies and the activities of bodies.
Chemical elements exist.

---

*If such formal presuppositions are species-specific—which means, according to
Skinner, that they are genetically constrained and thus "innate"—then why did it take
human beings so long to ferret them out? I think the reason has to do with the
development of writing. If the estimates are correct, human beings have been using
language for some 40,000 years and writing down their language for less than 6,000.
There are oral traditions that have come down in fragmented form from before the
time of written language, but there is nothing in the oral traditions that smacks of
conceptual self-examination nor even of the rudiments of logic. When we do examine
how we think, we pay close attention to the words and symbols that we record. This
is as true of the Greek thinkers of 300 B.C. as it is of modern analysts. One is led to
suppose, then, that questions about how this species thinks did not arise in any serious
way until we were capable of writing down our ideas and lines of argument for our
own later criticism and the systematic criticism of others.

Everything that exists is composed of earth, air, fire, and water.
Subatomic particles exist.

Ontological presuppositions may also be of a less general sort, such
as these:

Oxygen exists.
Skunks exist.
Quarks exist.

From the examples, you can see that ontological presuppositions
are likely to change over a period of time and that two people are
likely to disagree about ontological presuppositions (the mind-body
presupposition, for instance). But assertions like "The halogens—
fluorine, chlorine, bromine, iodine, and astatine—form binary salts
by direct combination with metals" can be viewed as true or false
statements only within a context of presuppositions that gives them
sense, including, among others, the presupposition that chemical
elements exist. To argue about such statements, to test them for
truth or falsity, and to draw inferences from them become possible
only within a context of ontological and other presuppositions.

Existence presuppositions are about individual things, whereas
ontological presuppositions are about kinds of thing. In general, we
presuppose that assertions about individual objects refer to identifiable existing objects. We can't discuss whether or not Harry Truman's eldest son was a Democrat because Harry Truman didn't have
a son. In literal contexts of assertion, we can't claim that a drugstore
that is a figment of the imagination either was or was not actually
robbed, as we saw in the example in Chapter 1. We do, of course,
talk about fictional, or hypothetical, or merely possible individuals,
and we do make perfect sense of such talk so long as it is clearly
understood that this is the ground on which assertions are being
made. We can assert that Sherlock Holmes used cocaine, understanding that we are talking about a fictional character, and we can
say that, *if* Harry Truman had had a son, he probably would have
been a loyal Democrat. But the contextual shift from making
assertions about actual individuals to making assertions about
fictional or possible individuals has to be clearly understood if we
are to carry on rational discourse.

Categorial presuppositions determine the properties we are willing to ascribe to things of given kinds. Human beings and other
higher animals are said to have properties associated with awareness

and emotions; inanimate objects in general do not have such properties. Typewriters may be in need of cleaning but not in need of psychotherapy. Human beings may be in need of both. Electrons and other particles below the molecular level do not have properties of temperature.[1]

Categorial presuppositions are particularly subject to change over a period of time as both scientific and common sense points of view change. Until the twentieth century, almost all western physical theory presupposed that physical objects have properties of absolute rest or motion. In the context of contemporary physical theory, rest and motion are relational properties: an object can only be said to be at rest or in motion relative to another object.

There will be more to say about categorial presuppositions later, when other distinctions have been introduced to mesh with the ones just made. Sometimes, of course, we violate categorial presuppositions deliberately, not to produce nonsense but to produce figures of speech that have a function other than the straightforward conveyance of information.

One of the features of language that makes it difficult to lay out distinctions such as these in a way that will stick is that we can constantly shift the level of a discussion from assertions about sticks and stones to assertions about *assertions* about sticks and stones or assertions about language itself. Now this is a complicated business, and I don't propose to draw you any further into a set of linguistic distinctions than I need to in order to set things up for a discussion of how and why scientific ideas differ from one time and place to another.

The rules that delimit contexts for making meaningful assertions—a set of presuppositions—can themselves be asserted in their own contexts or in the broader context of an explanation or negotiation about what the common ground for a discussion is to be. Presuppositions have a dual status then. They serve the semantic function of delineating contexts for meaningful assertions, and they themselves can be asserted within their own contexts. To complicate matters even more, they can, like any assertions, be logically abstracted as statements understood within their own contexts and assessed for truth or falsity.

Perhaps the best way to illustrate how presuppositions can be both semantic items and assertions is by analogy with definitions. In formal contexts, we often stipulate definitions of particular terms. In the Commonwealth of Pennsylvania, for example, the term "singleman" is defined for contractual purposes as "an adult male [human being] who has never been married." Definitions are not statements, but there are statements which reflect them and which are true in the contexts where the definition applies. So, in any legal context in Pennsylvania, the *statement* "A singleman is an adult male who has never been married" is a true statement. The definition makes the statement true. In another context, where the Pennsylvania definition does not apply—say, a singles bar in Tijuana—the statement might not be true.

Presuppositions are not definitions of terms. They are the rules that determine what does and what does not make sense in a given context, what *can* be said to be true or false. But like definitions they can be expressed as statements which are always true *within* the contexts they govern.

Suppose you are discussing the motion of objects within the context of Newtonian physics. There, as I noted earlier, it is presupposed that objects are either in motion or at rest in an absolute (rather than relative) sense. You might assert that a given object, *x*, is at rest at a given time. Within the context, one can determine logically what the statement "*x* is at rest at time *t*" implies and determine whether the statement is true or false, consistent or inconsistent, and so on. Further, within this same context, the general statement "Objects are either in motion or at rest" is *always* true, because it simply states one of the presuppositions that characterize the context.* But within the context of contemporary physics, where motion and rest are understood as relational, the statement "Objects are either in motion or at rest" does not turn out to be always true; it requires some hedging in order for us to assess it at all.

To recall the analogy with definitions, if the term "singleman" is understood in a given context as "any unaccompanied adult male," then within that context the statement "A singleman is an adult male who has never been married" is false, even though it is true by definition in the context of legal definitions in the Commonwealth of Pennsylvania.

What I identified above as *formal* presuppositions are those which have been taken by philosophers and psychologists to be universal— common to all human beings and perhaps rooted in the very

---

*Jay Rosenberg has pointed out to me that this account of presuppositions asserted within their own contexts is rather like Immanuel Kant's doctrine of synthetic *a priori* judgments. So be it.

physiology of the human brain. There may be some argument as to which presuppositions belong on this list, but it seems beyond question that there are such presuppositions and that this list is a fair statement of the most important ones. With the exception of the presupposition—"No object both has and fails to have a given property"—the statements which express formal presuppositions are not generally taken to be "logical truths," but they do delineate the boundaries of common sense experience, whether or not it ever occurs to a given person that they do. While there isn't much that would count as argument for the formal presuppositions, it is possible to explain at least some of them, that is, to explain why they *are* basic to human experience. A behavioral psychologist would argue that human beings, and other species as well, consistently behave in such a way that it is taken for granted that every event has a cause. Failure to behave thus, when the event is a perceived sound or motion, for example, would make it less likely that an individual would be prepared to flee from danger or defend itself or to take an opportunity for nourishment. In evolutionary terms, there is over-whelming selective pressure in favor of those individuals who consistently behave as if every event does have a cause and against those who do not.[2] That is precisely how species-specific or "innate" behaviors come to be universal.

The remaining kinds of presupposition—ontological, existence, and categorial—are clearly not species-specific, and they vary greatly from one cline to another. Moreover, as I suggested in Chapter 1, we can voluntarily drop one set of presuppositions and adopt another on either a temporary basis, as when we converse with a child, or a permanent one, as when we are "converted" to a new point of view.

Coherent Beliefs and Plausible Hypotheses

In 1632, when Galileo said "The earth moves," Pope Urban VIII said, roughly, "No it doesn't." That is what logicians call contradic-tion. Flat out, head-to-head contradiction. One of them had to be right, and the other had to be wrong. Now you might want to object immediately that, from the point of view of the theory of relativity, they were either both right, or both wrong, depending upon how you reinterpret what they said. A modern astronomer has to agree with both of them, in a way. When the astronomer describes the universe, he is likely to say that everything, including the earth,

moves relative to everything else. But when he makes the calculations to aim his telescope at some particular area of the night sky, he
has to think in Urban's terms or at least use tables in his calculations
that are designed from Urban's point of view.

But relative motion wasn't what Galileo and Urban were at odds
about. In the context of their disagreement, there was no way
around the confrontation. They weren't talking about the motion of
one thing relative to another but of absolute motion, as we would
call it now, in what they took to be well-understood absolute space.
There was no way that a third party could reconcile the difference
by saying that it all depended upon how you look at motion, that
position and movement could only be discussed as a relationship
between objects. The context of the disagreement presupposed that
motion was absolute. Either the earth moved or it didn't. You could
not consistently believe both. If you had tried to mediate the
argument by saying something like "Well, the earth both moves and
doesn't move," as the introduction to an explanation about the
modern conception of relative motion, you would have been
dismissed immediately as an incoherent babbler. On purely logical
grounds, you simply can't assert both in one context, and to do so is
to say something that is logically impossible. Understood literally,
without the hedges of a new physical conception, you violate the
rules of rational human discourse when you say that a given object
both has and fails to have a given property.

The minimal thing that we must require of a set of coherent
beliefs is that they be consistent. On logical grounds, two statements
in the same context are said to be consistent if there is nothing about
either of them that prevents the other's being true and inconsistent if
they are related in such a way that they could not both be true. In the
exchange between Galileo and Urban, the clash was direct: "Yes" on
the one hand, "No" on the other. The question of consistency is not
the historical question whether either one of them was correct in
modern terms but, rather, a logical question about the effect that the
truth or falsity of one statement would have on the truth or falsity of
the other.

Inconsistency does not always depend on such direct Yes-No
verbal confrontation or on putting the Yes and the No together in
one logically absurd assertion. It just as often depends upon how we
understand the concepts involved. For example, we all understand,
I think, that in order to pass a test for a driver's license, you have to

take the test. The point is so obvious as to require no explanation. So you couldn't consistently believe both of the following:

Schulz passed his driver's test.
Schulz never took a driver's test.

Even if Schulz had bribed the officials to say that he had passed a test he never took, that wouldn't give you a way of reconciling the two. You would probably put quotation marks around the word "passed" in order to underscore the anomaly, if you had to write the two sentences down. The two statements are inconsistent because of what we mean when we say that someone literally passed a test, and to assert both is to assert something that is conceptually impossible.

Concepts change. The modern concept of space is different from that of Galileo and Urban. Combinations of beliefs that are conceptually inconsistent in one context might become consistent if we revise our concepts. But at any given time there is a prevailing way of understanding things, and there are some cases where concepts are unlikely to change very much. Things can't be red without having color, for example, and one and the same individual can't be both a dog and a tree. Where there are logical or conceptual matters at stake, we often say that we "can't imagine" a given state of affairs. I can't picture a four-dimensional physical object, however hard I try, and I can't imagine one and the same individual being both a dog and a tree. Such things are conceptually impossible, as distinct from the blatant Yes-No cases that we identify as logically impossible.

There is a third sort of possibility-impossibility question that figures in evaluating our beliefs. A situation that we might perfectly well be able to picture or imagine is judged to be *theoretically* impossible if it is inconsistent with a general explanatory theory that we accept, even though logical and conceptual issues don't arise. It would be theoretically impossible for a new planetary body to enter the solar system without its motion being affected by the existing planets and the sun. Such a thing would be inconsistent with Newton's laws of motion and gravitation, even as they have been hedged and modified by contemporary astrophysicists. Wherever there is a well-established body of theory, we can generally make a clear determination as to whether or not a given supposition is consistent with the theory—whether or not it is theoretically possible.

But if we are going to try to separate coherent sets of beliefs from incoherent ones, we can't simply settle for saying that what we believe isn't impossible—that it isn't inconsistent from a logical, conceptual, or theoretical point of view. If we are willing to assert that something is the case and seriously expect someone else to believe it, we must be willing to defend our assertions as *plausible*—minimally worthy of belief. Plausibility requires something more than possibility.

Consider this as a hypothesis:

There are diamonds on Mars.

At this moment, I don't believe that there are diamonds on Mars or, for that matter, that there aren't. The question is, Is the hypothesis plausible? Might I come to believe it? Is it worth considering further?

Let's tackle the question of possibility first. The hypothesis is possible in all three ways mentioned: There is no logical inconsistency, and there is nothing in my understanding of either diamonds or the planet Mars that would prevent its being true. So far as I know, there isn't any well-established theoretical reason to doubt that there are diamonds on Mars.

Beyond possibility, there are some other, less formal, grounds on which to assess the plausibility of a hypothesis. First is the matter of precedent or instances: Do we know of a similar situation where such a thing has actually occurred? So far as this hypothesis is concerned, of course we do. The Earth is similar in many geological respects to Mars, and there are diamonds on Earth. Even if we couldn't explain *why* there are diamonds on Earth, we know *that* there are, and that precedent makes it more than just possible that there are diamonds on Mars. It confers a degree of plausibility to the supposition that there are.

A second question that affects the plausibility of a hypothesis is how it might enter into explanations. Would the hypothesis serve to explain a range of events that we don't have a satisfactory explanation for? Alternatively, could we explain the hypothesis itself? The diamonds-on-Mars hypothesis doesn't explain anything much, but we could explain how there came to be diamonds on Mars. We know how diamonds are formed on Earth, and Mars is similar enough to our planet that the chemical and physical processes involved in forming diamonds could well have taken place. So we can see how it could have come to be.

Having a precedent or an explanation lends plausibility to a hypothesis, but failure to have one or the other doesn't necessarily constitute grounds for saying that the hypothesis is implausible. There are some things that we know occur but can't explain (for example, how aspirin works). And there are some things that we are able to explain before we are sure that they happen. This is exactly the situation that physicists are in when they say, on the basis of their theories, that a given sort of subatomic particle is liberated in the interaction of other particles before any experimental evidence is brought in. It has happened frequently in the past fifteen years or so that the characteristic "tracks" of particles in a bubble chamber have been predicted by an explanatory theory before the tracks were observed.

There is a third consideration in assessing the plausibility of a hypothesis: How does it mesh with what we already believe? If a hypothesis is inconsistent with some of our beliefs, whether or not we are willing to consider it—whether or not it is judged to be plausible—will depend in large measure on how important the challenged beliefs are to us and how central they are to our overall pattern of beliefs. The diamonds-on-Mars hypothesis is consistent with everything I believe, including beliefs about both central and peripheral matters. If further exploration should produce evidence of diamonds on Mars, it wouldn't cause me to change any of my beliefs.

Here is a second hypothesis that I have heard people assert recently:

A human being can levitate (that is, can rise from the ground simply by wanting to).

I don't believe this one either. But, again, the question is whether or not it is plausible. Let's run through the criteria. First, the questions about possibility. On the face of it, there is no logical or conceptual difficulty. Is it theoretically possible? This comes down to the question whether or not levitation is inconsistent with a generally accepted explanatory theory, and in these terms levitation is not theoretically impossible. It is not inconsistent with any well-established physical theory to suppose that some unspecified agency, linked with a particular sort of emotional state ("wanting to"), might act on the human body in a given way. What bothers, of course, is that we know of no such agency. But to claim that levitation is theoretically impossible simply because we would have no theory to explain it is to suppose something about the present state of science that is simply false. There are a number of familiar events that cannot be explained by current physical theory, including events involving gravity. Partly for this reason, there is no theory about how gravity works that is agreed upon by the western scientific community. There are as many as eight competing explanations of gravity right now, but none of them is generally accepted.[3]

As a first step toward evaluating the plausibility of the levitation hypothesis, we have to say that it is possible logically, conceptually, and theoretically. There is no inconsistency. Now, on to the other three criteria for plausibility.

Is there a precedent for the levitation hypothesis? Do we know of any instance where it has happened, or do we know of similar events where the emotional state of a human being has produced such an extreme physical effect? Here the matter of precedent is less clear-cut than it was with the hypothesis about diamonds on Mars. The religious literature of both East and West contains accounts of incidents of levitation. How seriously do you take such accounts? Are they to be understood as literal eyewitness reports, or as metaphors of some sort to the effect that it was *as if* a given saint or other personage had risen from the ground? I don't think there would be as easy a consensus on this matter as there was about the precedents for the diamonds-on-Mars hypothesis.

But there are other apparent precedents for levitation. Photographic evidence has been published recently of American students learning meditative techniques, who—as the photographs seem to indicate—literally rose several feet from the ground while meditating.* In addition, there have been reports of related events involving "psychokinesis," where witnesses claim that objects can be moved about by a "pure act of will." Some of these reports are clearly fraudulent, but it isn't clear that all of them are.

So the matter of precedents for the levitation hypothesis is a bit muddy. Incidents have been claimed, but there is no general consensus as to the reliability of the claims. There have been fraudulent claims made in the past about similar matters, and it just isn't clear whether or not the recent claims are to be taken seriously.

---

*The photographs are published by the Students' International Meditation Society, which teaches a technique called transcendental meditation. There is every reason to believe that the photographs are from a reliable source. I have seen them, but I just don't know how to evaluate them.

As a public matter, the question of precedents is undecided. (If, on the other hand, you have witnessed or experienced levitation, the question of precedents is settled so far as your personal beliefs are concerned.)

Could we *explain* levitation? This is the second question related to the plausibility of the hypothesis. As I noted in considering the theoretical possibility of levitation, current physical theory could not explain it. Levitation is theoretically possible in the sense that it isn't inconsistent with any of our physical theories, but it gains nothing in plausibility from the physical theories because we could not explain it, even if we were confronted with indubitable evidence that it had taken place as described.

Finally, we come to the question whether or not the levitation hypothesis is consistent with what we already believe. Again, the matter is problematic. The relevant beliefs are precisely in an area where we lack a consensus. A belief that dominated western thought for hundreds of years is that "mind" and "matter" are to be sharply distinguished, that "mental events" and "physical events" influence each other only within the human body—where physical stimuli are linked to awareness, and where the decision to move a certain part of the body is linked to the actual movement. The levitation hypothesis would certainly challenge this belief. But the belief is already under challenge from other quarters. The sharp distinction between the mental and the physical that was until recently characteristic of western thought just doesn't seem to work very well in the practice of medicine, for example. Psychosomatic medicine, cures of obvious bodily ailments effected by hypnotic suggestion, and other developments have made it clear that the distinction needs to be reconsidered.

To ask whether or not a hypothesis is plausible is to ask for a critical examination of the hypothesis in the light of established beliefs and in the light of available evidence. To say that a hypothesis *is* plausible is to say that it is worth pursuing further by looking for more evidence that will support it or refute it.

We evaluate hypotheses both privately and publicly. Not all of us are equally critical concerning our personal beliefs. But scientific beliefs are another matter. A scientific community is a critical community in which hypotheses are publicly evaluated in rather stringent ways. For good reason, the scientific community tends to be more conservative about accepting hypotheses than most individuals are. Personal beliefs and scientific beliefs are not identical, then—not even when we are talking about the personal beliefs of a working scientist. Most working scientists have pet hypotheses that are not accepted by the scientific community at large; the public critical verdict isn't in yet.

Now, where do we stand on the two hypotheses we have been considering? The first, that there are diamonds on Mars, is quite plausible. It is surely possible in all the senses described, and the three additional criteria—precedent, explanation, and consistency with established beliefs—are all in its favor. It is clearly worth pursuing further, and no doubt there will be evidence gathered by an exploratory spacecraft that will have some bearing on the hypothesis.

The second hypothesis, that human beings can levitate, remains problematic. It is possible, in all three senses, but we didn't get a clear-cut answer as to whether or not it is something more than possible. There may be some instances, but it is hard to be sure. There isn't an available scientific explanation for it, at least not in contemporary physical theory. The beliefs relevant to it are precisely in that area where there isn't a good consensus at present.

But is the hypothesis plausible to you? That is a different question from whether or not it is scientifically plausible. Whether or not you or I take the claimed precedents seriously, they have not been established by controlled experiment, and that is what scientific plausibility demands. Whether or not you or I think that there might be some rather strong physical agencies that can be influenced by our thoughts, there is in western science no theory that has been subjected to critical scrutiny that would explain how such forces work.

I doubt that very many western physical scientists would grant the plausibility of the levitation hypothesis, at least not when "speaking professionally." Most would say that it is implausible, on the explanation grounds alone. We simply cannot see how it could happen. This may be too conservative an attitude for individuals to take. Perhaps the relative conservatism of the scientific community is too extreme to apply to our personal beliefs. On the other hand, some people will believe anything, and we must avoid that extreme as well.

Identifying, Counting, and Classifying

In Chapter 1 I introduced the question "How many objects in this room?" as a means of making the point that there is not a uniquely correct answer to such a question. There are ways of modifying the question that don't help at all. "How many red objects in this room?" doesn't do the job, for example; there is still more than one reasonable way to decide whether you have a single red object or a connected group of red objects. "How many soluble objects?" doesn't help either. "How many dogs?" on the other hand, admits of a straightforward answer. If you are counting dogs rather than things, there *is* one correct answer. Other terms, like "chairs," "soldering irons," "bricks," or "linoleum tiles" would work as well. They are all terms that tell us what kind of thing is being asked about.

I'm going to use the words "identify" and "classify" here as a biologist uses them: We *identify* individual things as being things of a kind; we *classify* a group of individual things (or all things of a given kind) as being within a more inclusive grouping. Individuals are never classified in this sense; they are identified as being of a given kind or as being composed of a given substance. Thus, we identify Fido as a dog, or a mammal, or a vertebrate, but we classify dogs as mammals and mammals as vertebrates.[4] The distinction can easily be extended to "stuff" or "matter" terms: We identify a given quantity of a substance as peanut butter, and we classify peanut butter as nourishing food.

There is not, of course, precisely one correct way of identifying objects; the activity can be carried out in a variety of ways, using a variety of different vocabularies. But the activity of identifying things is a very basic conceptual activity that has to do with the identity of individual things. To identify an object is to say what kind of thing it is, to delineate its boundaries in such a way that if it ceased to be that *kind* of thing it would also cease to be *that* thing.

Time for an example:

Fido was sick yesterday but is well today.

No problem. Being sick has no connection with Fido's identity, and it is just the sort of thing that comes and goes in a given individual. But how about this:

Fido was a dog yesterday but is a squirrel today.

It just doesn't work. If Fido stops being a dog, Fido ceases to be the individual he is.

There is nothing especially mysterious about basic identifications. They reflect not some arcane truths about the universe but, rather, the way we sort things out in a given context. Identifying a thing as a thing of a kind gives a basis for individuating it as a particular thing, for saying that it is one chair rather than twelve pieces of wood, for example. To play on the terms a bit, when we identify something, we give it an identity that allows us to count it and to describe it further.

But despite the close connection between identification and identity, we can't claim to have a unique way of identifying objects. There is more than one way of identifying and counting individual objects, both in common sense contexts and in the more systematic contexts of the sciences. There are situations where we voluntarily change our identifications—our basic sorting out of objects—for one reason or another, as when we consider how a cabinetmaker might identify the objects in the room as distinct from the way a moving man might identify them for an inventory. I will suggest shortly that re-identification of individuals and re-description of events figures importantly in certain kinds of scientific explanation.

It's the *activity* of identifying and classifying that is basic, not any particular scheme of identification or classification. Both identification and classification have characteristic patterns in natural speech, and these patterns are reflected more formally in the way the various sciences identify and classify the objects they study. In the next chapter I will take a brief analytic look at the way these activities are carried out formally in biology and chemistry.

Describing and Explaining

Suppose you are walking along a beach and you come on the remains of an exploded aerosol can. Some of the paint is burned off, and all the jagged edges stick out from the can rather than into it. So far, we are just describing what you've found, from what seems to be a reasonable point of view. We've identified the thing, and from its appearance it looks as if it exploded. If you start asking *why* the aerosol can exploded, you're asking for an explanation and inviting hypotheses.

A hypothesis, as I have used the term, is a supposition. Typically, when looking for an explanation, we try out a few hypotheses and sort out which of them is the most likely. If one hypothesis comes to be accepted, we will call it a *thesis*.

A plausible enough hypothesis about the aerosol can is that it was simply left in the sun, overheated, and exploded. If you can still read the relevant parts of the label, you might see a cautionary notice telling you not to leave the can in the sun. Another hypothesis is that the can was thrown into a fire while it was still sealed. This second hypothesis seems more likely than the first because of the burned paint. The next step is to take a look around the area ... sure enough, just over the dunes are the remains of a bonfire, giving even more reason to prefer the second hypothesis.

A straightforward causal explanation of this sort typically involves three items: a question that includes a description (why did the can explode?); one or more hypotheses; and a procedure for testing the hypothesis on its own, to see if it is plausible, independent of its connection with the event to be explained (the search for the site of the fire).

The next sort of question you might ask about the aerosol can is *Why* did it explode when it was thrown into the fire? The "why" questions can go on indefinitely. In this case, we have some well-established general theses about why such things happen. To explain why the can burst when it was thrown into the fire, you might talk about what happens when you increase the temperature of air (or any gas) while holding the volume constant: the pressure builds up until the container isn't strong enough to resist it, and the container ruptures.

Notice that there is a new step in this explanation that didn't occur in the first one. We have replaced the elements in the original description of the event as follows:

The closed container is re-identified as a system of constant *volume.*

Throwing the can into the fire is re-described as increasing the *temperature* of the system.

The explosion of the can is re-described as a symptom of an abrupt and extreme increase in the *pressure* of the system.

The first explanation of why the can exploded—because it was thrown into the fire—didn't require any re-description. Causal explanations typically do not. When we have to shift ground and re-describe an event in order to explain it, or to re-identify elements in the original description, the explanation becomes *theoretical.* By this I don't mean that there has to be an organized theory to back us up; I simply mean that there is a discernible shift in the language we use for the explanation that reflects a new context of presuppositions. It is exactly the re-identification of objects and processes that marks a shift between data (description and causal explanation) and *theory*.

The line between data and theory is not a fixed one, and it can be drawn only with respect to a given context. What signals that the line is being crossed in a given instance is the need to re-identify items and re-describe circumstances. But it isn't as if the original description was in some sense neutral or independent of its own context of presuppositions. We could as well have identified what was found as a jagged, torn piece of metal and then re-identified it as an exploded aerosol can. The data-theory line isn't fixed in either direction from what seems in a given instance to be a natural context of description. (This is what I take Karl Popper to mean when he says, "We are theorizing all the time."[5])

The explanatory re-description introduced the terms "volume," "temperature," and "pressure." But these same terms can serve to describe data for further theorizing. If you were to ask the more general question *why* the pressure of a system increases when the volume is held constant and the temperature is increased, you would most likely get an explanation in terms of the kinetic theory of gases. This involves a shift to yet another context of presuppositions, in which the following description can be given:

Temperature is re-identified as the mean kinetic energy of gas molecules.

Pressure is re-identified as the velocity with which gas molecules collide with the wall of the container.

The line between data and theory can be drawn on such contexts, but it cannot, in general, be drawn in any permanent way on a whole language, or even on the whole vocabulary of a given scientific discipline. In the first instance of theoretical explanation, the question was about an aerosol can thrown into a fire, and the answer was about temperature, pressure, and volume. In the second instance, the question was about temperature, pressure, and volume, and the answer was about the energy and velocity of gas molecules.

An explanation of an event (or kind of event) is *causal* when it invokes hypotheses about other events but does not necessitate re-describing the event to be explained. A hypothesis is typically a trial explanation. Suppose that the can had been thrown into the fire; that would have caused it to burst. Further checking (independent evidence) shows that there was indeed a fire nearby, and the explanation seems plausible.

An explanation of an event (or kind of event) is *theoretical* when the hypotheses or theses invoked necessitate re-describing the event to be explained. There is a locution that is often used in such cases that harks back to the argument about the red shirt in Chapter 1. It's this: "What *really* happened is that the mean kinetic energy of the gas molecules increased, so that they collided with the walls of the container with greater velocity than the metal could withstand." The word "really" signals a preferred way of describing the event, but it also suggests a kind of unique correctness that just isn't present.

Truth, Falsity, and Accurate Description

The position I have taken about asserting beliefs and describing situations provides a way of dealing with statements that occur within different contexts of presupposition without trapping us into the kind of vicious relativism that leads to dismissing questions of truth and falsity altogether. Too often we hear people suppose that what they say can be "true for me" whether or not it makes sense to anyone else. Such a supposition misses the most important feature of claims for truth and claims for knowledge: they are objective, not subjective, claims. The fact that there is not a *uniquely* correct way of describing a state of affairs should not mislead us into thinking that we can't sort out correct descriptions from incorrect ones.

We can judge statements to be true or false only if we understand the context in which they are asserted. The objects we identify are objects of our experience, and the properties they have are natural properties by virtue of human nature; they are a function of the relationship between human beings and the objects they attend to. Beyond this, there are indefinitely many ways of identifying objects and describing their properties—determined by language, local idiom, the purpose of the identification or description, and the overall context of presuppositions in which any discourse takes place.

But this is a rather confusing claim, because to challenge it you must start asking whether or not objects "really" are the kind of objects we take them to be and whether or not they "really" have the properties that we ascribe to them. Such a question involves a fundamental confusion about what is involved in making an objective claim that a given description is accurate or that a given statement is true. I hope to convince you of this before the end of this chapter.

Now, before I lay out a way of looking at judgments of truth and falsity, let me soften you up with an example. Figure 3-1 shows three different maps. The top map shows how land is used in the area near an electric power line. The middle map represents points in a given geographical area that are the same height above sea level. The bottom map is a street map of part of a city, as you might guess.

In a fairly obvious sense, each of the three maps may be said to be an accurate map or an inaccurate map of the area it represents, once we know how to interpret the symbols in each map. As a matter of fact, they are all accurate maps, within reasonable standards. Now, to be able to *tell* that they are accurate, you would have to know what each different sort of line and symbol on each map represents—what it corresponds to in the physical area that the map covers. In addition, you would need to understand how to determine whether or not the relevant features of the geographical area depicted in each map are indeed as they are represented. Without understanding these things, which are not part of the maps themselves, you would have no way of evaluating the maps.

Before I connect the example with assertions, there is one more thing about the three maps that should be noted now, so I can refer back to it later. The top and middle maps are recognizable as maps of the same territory. Without even knowing what city the maps represent, you can probably tell that these two are maps of the same area. But chances are that you can't tell from the maps themselves whether or not all three represent the same area, and, if you were to compare the bottom map with either of the others alone, you would not be able to tell whether or not these pairs represent the same area. As you've probably guessed by now, all three maps do represent precisely the same area (part of San Francisco). But this is clear from the maps themselves only in the case of the first two.

There are several points to be noted about the maps:

Each map stands in its own relation to the geographical area
   it represents. There is no "basic map" that must stand
   between each of the other maps and the geographical area.
Each map can be understood and assessed for accuracy only
   if its *legend*—the idiom of its symbols—is understood.

Figure 3-1 • Three Maps

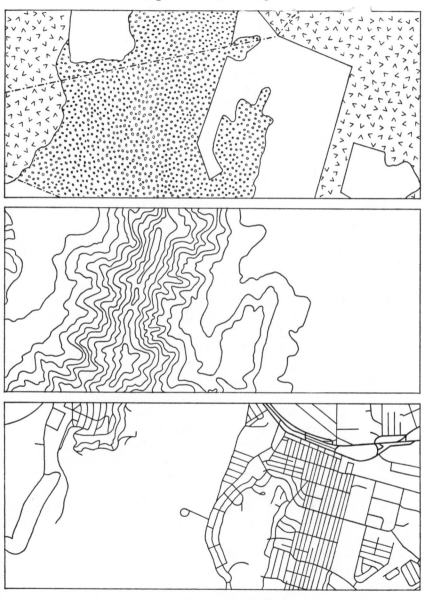

Cartography by Janet M. Wallis

The elements of each map either do or do not correspond to
discernible elements in the geographical area represented.
The question is a purely objective one, which is to say that
anyone who understands the idiom of a given map and has
access to the area it represents can determine the extent to
which the correspondence exists.

The relationship between each map and its geographical area is
independent of any *particular* individual's beliefs about it.

No one of the maps is in any sense *uniquely* correct. All of them
are accurate, and a preference for one over the others must
have as its basis something other than the question of accuracy.

Truth and falsity are properties of statements in the view I have
taken here. We assert our beliefs within well-understood contexts,
and our assertions can be assessed as true or false statements only
within their own contexts.

If I describe the facts of a given situation, you can determine
whether or not my description is accurate, provided we are on
common ground—provided, that is, that you understand the context
in which I am offering the description. An accurate description does
not carry with it the guarantee that it is the only accurate description. Someone else might describe the same situation accurately
from another point of view.

The point of view that is characterized by a given set of presuppositions does not decide the facts; it merely provides a way of
describing them in a set of statements. Whether or not the statements are true is a matter for empirical investigation. The whole
descriptive package of presuppositions and statements is then
subject to further judgment concerning the adequacy of the account
it gives, relative to the reasons why the description was undertaken
in the first place. What I am considering here are, of course, ways of
evaluating descriptions of particular situations. I will discuss beliefs
of a more general nature in the next two chapters.

We already have an example before us that provides several
descriptions of the same event: the explosion of the aerosol can on
the beach.

In the first description we simply assert that at a given time, t, and
at a given place, p, an aerosol can was thrown into a fire and that it
exploded.

In the second description, also about what happened at time t and
place p, we assert that a container of gas at constant volume was

subjected to an increase in temperature, so the pressure increased sufficiently to rupture the container.

In the third description we assert that at t and p the mean kinetic energy of a system of gas molecules was increased, so the velocity with which the molecules struck the walls of the container exceeded the breaking strength of the walls.

Given the example, all three of these alternative descriptions are accurate. Each of them corresponds to the facts under a given interpretation, which is just to say that the incident can be accurately described in any of the three ways. Assessing the accuracy of each description entails understanding the context in which it is asserted. Alternative descriptions need not have any more in common than that they all are connected with the world in a recognizable way by their own semantic relationships. It need not be the case (as it is in the example) that every verbal element in one set of statements can be matched up with a verbal element in the others.

Now, some observations concerning the alternative descriptions which parallel the earlier observations about the alternative maps:

> Each description stands in its own relation to the facts. There is no "basic description" that must stand between each of the other descriptions and the facts.
>
> Each description can be assessed for accuracy or inaccuracy only if the context of presuppositions in which it occurs is understood.
>
> The statements in each description either do or do not correspond to discernible elements of the incident described. The question is purely objective, which is to say that anyone who understands the idiom of a given description and knows what it is about can determine whether or not it is accurate.
>
> The relationship between the description and the incident is independent of any particular individual's beliefs about it.
>
> No one of the descriptions is *uniquely* correct. All of them are accurate, and a preference for one over the others must have as its basis something other than the question of accuracy or inaccuracy.

The *objectivity* of claims for the accuracy or inaccuracy of descriptions is a matter of intersubjective agreement: a set of descriptive statements asserted within a given context of presuppositions has the objective property of truth or falsity if anyone who understands the context of presuppositions and knows what the statements are about can determine whether or not the statements are true. Objectivity, then, does not entail unique correctness. The statements in each of the three contexts are true. Each of the three descriptions is accurate. Which of them tells what *really* happened on the beach? They all do.

FURTHER READINGS

This chapter reflects a good deal of recent critical dialogue among philosophers. For general discussion of the relevant issues, see Stephen Toulmin, *Knowing and Acting* (1976), and Jay Rosenberg, *Linguistic Representation* (1974).

On presupposition, the seminal work in this field is in P. F. Strawson, *Introduction to Logical Theory* (1952). The present account differs from Strawson's in treating presuppositions as both statements and semantic rules.

On possibility and plausibility, for a more technical development of the concepts, see D. Paul Snyder, *Modal Logic and its Applications* (1971), Chap. 6. For a formal treatment of existence presuppositions, see Chaps. 5 and 8.

On truth and objectivity, see Karl Popper, "A Realist View of Logic, Physics, and History," reprinted as Chap. 8 of his *Objective Knowledge* (1972). The view presented here is very close to Popper's.

Readings on the theory-data distinction and on explanation are listed at the end of Chapter 4.

FOUR
Science as an Organized Activity

Science as we recognize it in western civilization today is a highly organized group activity. There is an identifiable scientific community that now crosses national and linguistic lines and that performs its specialized behavior on behalf of the entire community. Scientists did not evolve separately from the rest of humanity, of course, any more than teachers, dancers, cooks, farmers, whores, or handymen did. All parents, for example, teach their own children, but some people specialize in teaching children and do it on behalf of the whole community.

Similarly, everyone wonders about why things happen and develops hypotheses to explain why things are as they are. Everyone develops and acts on generalizations about what it is reasonable to expect in given situations. Scientists specialize in these activities and carry them out in systematic ways. Such division of specialized labor is important to the organization of human communities, and there are, of course, analogues to be found in the communities of other "social animals."

My emphasis on the verbal character of the activities of identifying objects and describing and explaining events may seem excessive on the face of it. But it is precisely the communication of scientific ideas in a systematic way that makes it possible for science to function as a public activity and that provides the scientific community itself with the critical devices that sort out acceptable scientific judgments from unacceptable ones. Such judgments are always subject to critical evaluation within a given scientific discipline and within the context of presuppositions that characterizes the idiom of a given discipline at a given time.

The objectivity that science requires depends in large measure upon the public examination of specific assertions in the light of evidence and in the light of already accepted theoretical judgments. If there is a "scientific attitude" that can be encapsulated in a slogan, it is this: *Assert only what you can defend.*

The awe with which we tend to treat scientific assertions leads some people to suppose that the slogan should read: "Assert only what you can prove." This is misleading, because there are at least two distinct uses of the term "prove" that can cause confusion about just what it is a scientist does. On the one hand, there is the sense of proof within a mathematical, geometric, or logical system wherein a given result is derived from a set of axioms and definitions by means of specific inferential rules. Given the logical rules, the axioms, and the definitions, the result follows by necessity. Scientific proof, on the other hand, reflects another long-established use of the term that is closer to the German word *Prüfung*—"testing."[1] This use is reflected in the notion of proving a weapon (as a sword, or the more modern weapons that are tested at military "proving grounds") or the proofs that are struck from a page of type. It is also reflected in the expression, "The exception proves the rule," which is absurd on the mathematical proof understanding. The apparent exception to a rule *tests* the rule and its ability to account for its subject matter. It doesn't, of course, demonstrate the rule. And, finally, how do you prove a pudding? If we want to rephrase the slogan, it might read: *Assert only what can be publicly tested.*

I have already described this book as containing a theory about science. It is, in that sense, a description of what happens when people do science; it is not a prescription about what *should* happen, except in the sense that it might underscore certain methodologies which have proved successful. I believe that a philosopher of science should study scientific methodology and theory in much the same way that a linguist studies the grammar and content of natural language—as an empirical study of a living system. Any pronouncements about what it is to do science should be approached with the same critical attitude—the same demand for public testing in the light of evidence—that pronouncements in the physical sciences receive.

Data, Theory, and Nomenclature

Many of the key terms that are used to describe scientific activity have suffered from overuse and have taken on a variety of different meanings. "Theory," for example, might be used to refer to Einstein's theory of special relativity—a well-articulated physical theory—or your maiden aunt's theory that her bathtub is haunted—an unconfirmed belief, at best—or any of the many theories about the assassination of President Kennedy—conjectures that are consistent with the known data. I think it best to set out here just which of the various uses of key terms I am adopting and to be as scrupulous as possible hereafter in using them always in the same way.

Data

The word "datum," of which "data" is the Latin plural, means simply "fact" or "what is given." It is customary to use this plural term in singular constructions. The data, collectively, is what happens—what we experience, in general. But, because there is no singular way of describing our experience, there is no clear-cut segment of language that we can set apart as the vocabulary for describing data.[2] Recall the example of the aerosol can in the last chapter. In the first instance, the data consisted of a description of a container that had burst when it was thrown into a fire, and what was introduced to explain the data was a theoretical re-description of the situation in terms of temperature, pressure, and volume. In the second instance, temperature, pressure, and volume entered into the description of the data, and the theoretical re-description was in terms of molecules of a gas and their properties.

Whether a given term enters into a description of the data or into a theoretical explanation of the data will vary with context. The line between theory and data often gets lost in practice and emerges as a clean line of distinction only when a difficulty arises with a given theory or when an analysis of the theory is carried out.

The data level for an emerging theory is the level at which no one concerned would disagree about the accuracy of descriptions. In such a situation, the data must be described at a "ground level" of general agreement, so that even rival theorists will be able to agree on it.

The data level of a well-established theory will be different from that of a theory that is still in the hypothetical stages and will be more influenced by the theory itself. Statements expressed in the data vocabulary describe what the theory explains, and the vocabulary is often constrained by what the theory can explain. An orthodox style of description develops relevant to a given well-accepted theory (as in contemporary chemistry and biology), and it is likely to be abandoned in favor of a more basic style of description only when the theory comes under attack. So the question whether or not a given term is part of the data vocabulary is not a logical or even a linguistic one. It is an empirical question, which is to be answered in a given instance by an empirical study of scientists at work.

We can, in a given context, talk about data statements and vocabulary as distinct from other portions of the context; the distinction is made according to how terms function in that context. Specifically, in general chemistry, the data vocabulary is used to describe what is measured and observed in chemical reactions—the colors, textures, weights, and volumes of substances; the temperatures at which they undergo physical change, such as boiling, freezing, or melting; what visible changes happen when we drop pellets of zinc into hydrochloric acid or pour sulfuric acid over sugar, and so on. In zoology, the data is described in what is called the *phenetic* vocabulary—the general appearance of organisms, their skeletal structure, internal organs, coloration, nourishment, breeding patterns, and so on.

Hypotheses and Theories

A hypothesis is offered in an attempt to explain data. Hypotheses that require a re-identification of the items to be explained I will call *theoretical hypotheses*. The ancient supposition that the lights in the sky are holes in concentric spheres which surround the earth is just such a hypothesis. Explanatory hypotheses that do not require such re-identification will be called *data-level hypotheses*. Where context makes it clear which is meant, I will use the term "hypothesis" alone.

Data-level hypotheses typically figure in causal explanations. When you explain a broken cup on the kitchen floor by supposing that the dog knocked it off the table, there is no need to re-describe the situation you are explaining. The hypothesis is at the "same level" of discourse as the description of the data to be explained, which is to say that the description and the hypothesis occur within the same context of presuppositions.

An explanatory hypothesis which is theoretical in the sense just described is not necessarily part of a well-articulated theory. Some theoretical hypotheses never get beyond the stage of supposition or the suggestion that we "look at it this way." They are eventually discarded without ever being taken seriously enough to enter into an organized theory. Other theoretical hypotheses survive, either by standing up to critical examination better than competing hypotheses or by receiving confirmation from further independent data. These may become *theses* of a nascent theory. A group of related theses may finally be linked together in a fully developed theory. I will reserve the term *theory* for such organized groups of theses.

Each of these terms has to do with the role that a statement plays in a given context and at a given time. A statement which begins as a theoretical hypothesis may become so well established that we are willing to call it a thesis; it may then be integrated with other statements into a theory. Over a period of time the theory may become so well absorbed into familiar discourse that it becomes commonplace—part of the everyday vocabulary used to describe experience directly. Theses about the temperature, pressure, and volume of gases have certainly been thus absorbed.° In such cases, a statement which begins as a theoretical hypothesis ultimately becomes part of a straightforward description of data and is itself subject to further theoretical explanation.

Moreover, the line between what we experience and what we do not is blurred in several ways. How sophisticated our concepts are and how we connect them in immediate, non-inferential ways to sensory stimulation varies from one person to another. An experienced mechanic listening to an automobile engine literally hears sticking valves and loose tappets, whereas the owner of the car hears knocking and pinging noises. Our technology also blurs the line. We are able, in some cases, to translate stimuli that are outside the range of the senses into stimuli that are well within sensory range. We see—in the direct sensory process—differences in color. We do not see differences in infrared (heat) radiation or in the reflected radiation from other parts of the electromagnetic spectrum. But we can translate reflected heat or other invisible radiation into color images through a highly developed technology for doing so. Other sorts of property can be brought within the sensory range by means of gauges or other measures or by electronic visual display. In well-

°In some circles the vocabulary of a given psychological theory is often used, inappropriately I suggest, to describe directly the way people behave.

understood contexts, such information unquestionably serves the function of data.

Theory, on the other hand, has an explanatory function in a given context. In contemporary chemistry, for example, the theory that explains why substances undergo the changes that they do is expressed as a series of statements about the atomic structure of chemical elements, their atomic numbers, the character of their nuclei, the arrangement of shells or rings of electrons around the nucleus, and the bonds that form when chemical elements are joined in compounds.

In zoology, the theory is expressed in a series of statements about the evolutionary interrelations among species of organisms. The *phylogenetic* vocabulary describes the branching lineages of the descent of species from other species. Certain physical resemblances among populations are identified as homologous—due to inheritance from common ancestry—and lineages are traced from one species to another, sometimes including species that are extinct and whose fossil remains may or may not be available.

Nomenclature

An important part of both description and explanation is the identification and naming of things under study. In the preceding chapter I called attention to the distinction between identification and classification: an individual thing, or an individual sample of matter, is identified as being a thing of a kind in such a way that the identification is linked to its being distinguished as the particular thing that it is. Kinds of thing, in turn, are classified in more and more general ways, in a hierarchical arrangement that may or may not be very strict.

In chemistry, the nomenclature consists of the names of chemical elements and compounds and of hierarchical groupings of these names that are rather loose (as compared with biology, for example), such as halogens, inert gases, and metals. Kinds of stuff are classified: the element iodine is classified as belonging to the "family" of halogens, and halogens in turn are classified as non-metals.

The identification of the chemical substance of which something is composed is bound up with the thing's being the particular thing it is, in much the same way that the identification of a thing as being a thing of a given kind is. It is a strong trait of at least all western languages that individuality is of things. *This* crystal of salt, or *this* chunk of copper sulfate, ceases to be this individual thing when it is dissolved in a liquid or recombined in a chemical reaction. A given sample of matter may be identified as a non-metal, or as a halogen, or as iodine. The last identification is of course the most satisfactory of the three, because it is the most specific. It identifies the sample at the least general of the three levels, and thereby connects the identification with a set of properties associated with iodine, and a set of theoretically based judgments and expectations concerning the sample in question.

In biology, nomenclature consists of the names of taxa, and these are arranged in a highly structured way. A taxon is a group of organisms recognized as a formal unit in systematic taxonomy, and taxa are ranked from the basic level *species* through the more inclusive levels *genus, family, order, cohort, class, phylum,* and *kingdom*, with intermediate levels sometimes distinguished (such as superorder and suborder) between the ones mentioned. A given species of dolphin will be further classified in phylogenetic taxonomy as of the genus *phocaena*, family *delphinidae*, order *cetacea*, cohort *ungulata*, class *mammal*, subphylum *vertebrata*, kingdom *animal*. Classification of a given population must take place at all the major levels mentioned to be systematically complete.[3]

An individual may be grossly identified as an animal, a vertebrate, or a mammal, but as was the case with chemical identification, the biological identification of an individual is more satisfactory as it gets to the least general level. An individual specimen is completely identified from a biologist's point of view when its species is given.

Again, it isn't just a matter of neatness or compulsive thoroughness that makes complete identification preferable. The more specific the identification, the more we know what to expect of the item in question and the more we know how it is connected with other items.

General Patterns of Activity

The three activities of describing, explaining, and identifying are interrelated. In the informal cases discussed in the preceding chapter, the relationships are blurred, but in the formal extensions of those activities that comprise the organized sciences we can see the lines of connection more sharply. Typical patterns emerge in chemistry and zoology, and they can be found in the other organized sciences as well. The general pattern of relationships among the three activities can be laid out as follows:

Figure 4-1 • The General Pattern

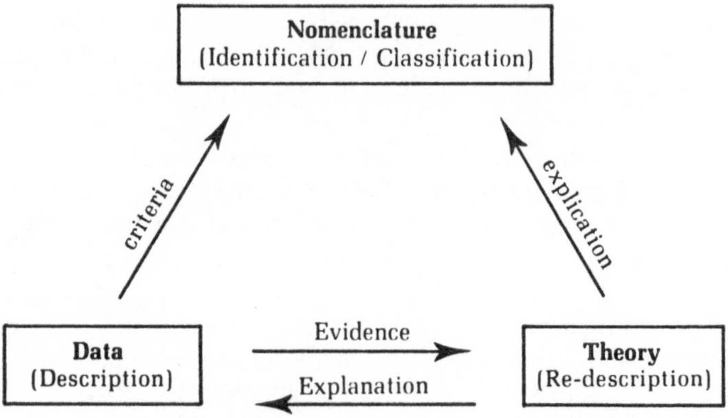

Criteria and Explication

Nomenclature in the sciences is a systematic activity. The names of items under study are used in both description and explanation: In general, data statements about things of kind K will tell what characteristics Ks have; theory statements about Ks will tell why they have the characteristics that they have. But it is the theory, not the data, that provides the systematic scheme for classification.

We cannot classify on the basis of all possible combinations of descriptive characters. The classification becomes impractical, for one thing; there will be many classes which are describable, but empty. There is nothing in the mere listing of characters that gives a clue as to whether or not there might be some undiscovered kind of object that meets the description. Systematic classification is done on the basis of the best available theory which explains why things have the characteristics they do. This means that classificatory schemes will change from time to time, as the best available theory in a given field is modified or replaced.

The advantage of classifying on the basis of theory rather than data is considerable. Either way there are going to be some classificatory slots which have no known members. But although a classification of chemicals based on data will have a place for a purple gas that liquefies at 4° Centigrade and produces an explosive reaction on contact with metals, it will give us no way of telling whether or not there is such a gas—not even good reason to believe that there might be. Classification on the basis of explanatory theory, on the other hand, gives good reason to believe that there are certain kinds of thing that have not yet been discovered. Such a classification can often function to predict that certain chemical substances might be found or fabricated or that certain biological organisms must exist (or must have existed at one time) in the genetic chain that links the known species. In many cases, such predictions have been borne out by the laboratory creation or the discovery of new elements, or the discovery of fossils or even live specimens of previously unknown species.

Chemical elements are classified on the basis of the periodic law, which states roughly that the properties of the elements vary in a periodic fashion with an increase in their atomic number. The variation is explained by reference to the electronic structure of the atoms of the elements based on the Bohr model of the atom (which is much simpler than the conception of the atom in contemporary physics, but adequate for chemical theory).

The modern classification of the elements is usually credited to Dmitri Ivanovich Mendelejeff (1834-1907), a Russian scientist, who worked out a periodic table of the elements in 1868 based on his version of the periodic law (which used atomic weights instead of atomic numbers). The table had a number of unfilled spaces in it; there were no known elements to fill the spaces, but the theory demanded that there be such elements, and because of the periodic law, the theory allowed a fairly detailed description of what properties the unknown elements must have.

The end of Mendelejeff's story should come as no surprise. In 1875 the element gallium was discovered, with just the properties that the Mendelejeff classification had predicted for it. And in 1879 gallium was followed by scandium, and the theory gained general acceptance. By the 1890s the periodic table began showing up in textbooks, even though there were many blanks yet to be filled; for example, germanium, the inert gases helium, neon, and argon, and the "rare earths" all remained to be filled in. It still happens from time to time that a new element is created in the atomic laboratory which has no known occurrence in nature (the unstable halogen astatine is an example of this, as are the radioactive elements which have received more publicity), or that a laboratory element is discovered in small quantities in nature.[4]

In biology, the relevant theory is of course the theory of natural selection that was discussed in Chapter 2. The theory is of a different sort than chemical theory, and classifying on the basis of it requires a "reconstruction" of the phylogeny—the genetic relationships among organisms—from physiological data. Although the method of the reconstruction needn't concern us here, it involves sorting out characters; those which result from phylogeny must be sorted out from those which result from adaptation. A whale's general shape, for example, would place him closer to fish than to mammals, but the shape is taken to be part of the whale's adaptation to an aquatic habitat, and his mammalian characteristics are taken to be part of his phylogenetic heritage.[5]

Knowledge of the story of evolution is inferential and incomplete, and the classification of organisms is not nearly as neat and compact as the classification of chemical elements. But despite frequent revisions necessitated by the discovery of unexpected fossil remains, the theory has been able to provide a basis for classification which, like chemical classification, allows for the "prediction" of species that must have existed between known species. From time to time, we hear of the discovery of another missing link in the phylogenetic chain that has just the characteristics that the theory leads one to expect.

We can't leave this topic without noting the difference between the two sorts of explanatory theory just described and noting some recent developments relevant to each. Biological theory has a time dimension; chemical theory does not. There are *historical* gaps in the biological theory that may never be filled in. Interestingly enough, astrophysicists have recently developed hypotheses about the *evolution* of chemical elements, a process which takes place within certain kinds of stars, in which simple particles are combined into simple atoms and these in turn, through a process of fusion at certain times in the life-cycle of a star, are brought together to form the heavier elements. This new theoretical development will probably have no effect on the chemical classification of elements, although it does explain how the natural chemical elements came to be.

On the other side of the coin, there has been great excitement among biologists in recent years about molecular biology—the study of the genetic chemicals DNA and RNA—which promises to explain the mechanism by which a living cell replicates and *how* mutations occur in the structure of the DNA and RNA molecules. (Evolutionary theory needs only to claim *that* mutations occur.) It is not too farfetched to suppose that within the next fifty years or so molecular biology will have progressed to the stage where a full analytic "map" of a given DNA molecule can be drawn, and connections can be established between a particular DNA structure and the conspicuous characteristics of the organism whose cells have that structure.

Once both of these things have been accomplished, a new basis for biological classification will be at hand: organisms will be classified on the basis of the *new* best available theory which explains why they have the characters they do, and at the same time explains how the evolutionary process took place.[6] One would expect a DNA-based taxonomy to mesh rather well with existing evolutionary taxonomy, perhaps correcting some of the inferences that have been made about particular lines of ancestry but supporting the overall taxonomic structure.

Although theory typically provides the schematic basis for classifying things into kinds, the data—the descriptive statements— provide the *criteria* for deciding whether or not a given individual is of a given kind. Zoologists use a "key" in the field for identifying individual specimens by their conspicuous bodily characteristics.* In chemistry, the criteria for deciding what a given substance is consist of a series of laboratory procedures or tests, typically using known chemicals in reactions with the unknown one to make the identification.

There is a term that is conspicuous by its absence here: *definition.* There are a number of different ideas as to just what it is to define a term, but what they come down to is this: To define a term is to set the limits for its correct use, and this can be done in a number of different ways. Providing a precise synonym is probably the least common way, because it is seldom possible to provide a second term that will be synonymous with the first in all contexts.[7]

---

*There are field guides to birds, animals, and plants which are just such keys. Keys are typically hierarchical, like systematic theoretical classifications, but the hierarchy is one of more or less conspicuous characteristics by which a specimen can be recognized, and the names given are of course taken from the theoretical hierarchy and arranged along different lines. A key is a systematic listing of known species, used for identification only.

We can say that the term "halogen" is properly applied to all and only the chemical elements fluorine, chlorine, bromine, iodine, and astatine, and we have defined the term in one way. We can say that the halogens are chemical elements which have seven electrons in the outer shell, have a chemical valence of minus one, and occur just before the inert gases in the periodic table, and we have defined it in another way. Or we can say that the halogens have relatively low melting and boiling points, react vigorously with metals to form binary salts by direct combination, have such-and-such colors, odors, and standard states, and so on, perhaps even outlining a series of laboratory tests to determine whether or not a given bit of gaseous, liquid, or solid matter is a halogen or a compound of halogens, thus defining the term in yet a third way. Any of these three methods, and perhaps others, is a legitimate way of defining the term that must be kept distinct for our purposes here, so none of them will be singled out as the "correct" way of defining a term in the nomenclature.

For convenience, I will use the term *explication* for the theoretical scheme of classification, including of course the theoretical statements that tell what it is to be a thing of a given kind. The data statements, which describe things of a given kind, provide the *criteria* for deciding whether or not a given specimen is of that kind.

What we have now are a set of functional distinctions among the terms used in the sciences: theory, data, and nomenclature are sorted out according to how the terms and statements of each are used. As with any set of distinctions that are made on a living system, there will be places where the lines are difficult to draw—particularly the line between theory and data. But in a given context, some terms will clearly be data terms, functioning solely to describe, others will clearly be theoretical terms, functioning solely to explain, and others will be terms of the nomenclature, used to identify and classify and occurring in statements of both data and theory.

Theory explicates the nomenclature; data gives criteria for applying the nomenclature. What remains to be considered are the lines which connect theory and data. Their character is of course implicit in the way I have made the distinction between theory and data. The theory explains the data, and, to the extent that the explanation is successful, the data provides evidence and support for the theory.

Before we move on to the character of theoretical explanation and the question of how theories are supported by data, here are the patterns of Figure 4-1 as they apply to chemistry and biology:

Figure 4-2 • The Pattern in Chemistry

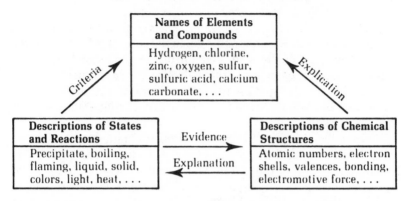

Figure 4-3 • The Pattern in Biology

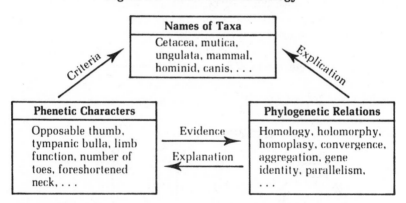

Theoretical Explanation and Prediction

Now we get down to a good thorny philosophical question: What is it to explain something? The sort of explanation described in the preceding chapter, which entailed re-describing the data in order to explain it, is what I take to be a typical theoretical explanation. Perhaps the best way to analyze the pattern further is with a simple example.

One of the first laboratory demonstrations in an introductory chemistry course is the hydrogen generator. Scraps of zinc are placed in a laboratory bottle and dilute sulfuric acid is poured into the bottle. The gas is collected in another bottle, which is inverted in water (to seal it from the air) and connected to the first with glass tubing. The result is a flammable gas in the collection bottle, and the zinc disappears in the acid solution in the generator bottle. The presence of hydrogen gas in the collection bottle is usually demonstrated by simply igniting the gas as it meets the air when the collection bottle is removed from the water. The laboratory setup is shown in Figure 4-4.

In practice, there is, of course, no problem in identifying the zinc and the sulfuric acid that are used in the chemistry lab. They come neatly labeled from the supply house, and we assume, safely, that the appropriate identifying tests have been made. If there should be a doubt, there are reasonably straightforward criteria and tests that can be used to assure that what we have are indeed zinc and sulfuric acid. Zinc is a bluish-white metal, malleable at 100–150° Centigrade and brittle above and below that range. The acid is tested by its reaction with other known chemicals.

Figure 4-4 • A Simple Hydrogen Generator

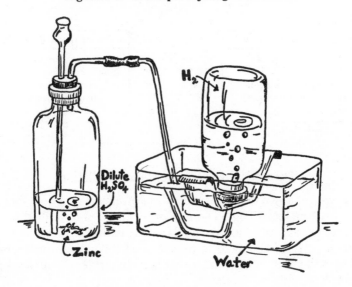

Drawing by L. Dogan, after Pauling.

Now, to explain what happened in the lab, the chemist first introduces a formula containing abbreviations of the names of chemicals:

$$H_2SO_4 + Zn \longrightarrow ZnSO_4 + H_2$$

Sulfuric acid, combined with zinc, yields zinc sulfate and hydrogen gas. Why? Here the theory begins. The experiment is re-described in the context of the presuppositions of chemical theory without contradicting the earlier description of what happened in the lab.

The atoms of the elements are understood on the Bohr model as consisting of nuclei with "shells" of electrons surrounding them. The number of electrons in the outer shell of each atom determines its *valence*—its combining capacity with other atoms—expressed as a positive or negative number. In a given compound, the valences must, in general, "balance out" to zero.

Zinc is understood to be an active metal with two electrons in the outer (fourth) shell of its atom. Its chemical valence is +2. Sulfuric acid is a compound of two hydrogen atoms and the sulfate radical (which is in turn compounded of a sulfur atom and four oxygen atoms). Hydrogen has a valence of +1. The sulfate radical has a valence of −2 and is thus combined with two atoms of hydrogen to form the acid.

Because of the electronic structure of its atom, zinc is higher in electromotive force than hydrogen, which is to say that it easily displaces hydrogen from acids. When zinc is combined with sulfuric acid, each atom of zinc (+2) bonds with the sulfate radical (−2), forming zinc sulfate, which is an extremely soluble salt. (Thus the "disappearance" of the zinc in the demonstration: the zinc sulfate was dissolved in the water which diluted the acid.) Two atoms of hydrogen (+1) are displaced by each atom of zinc (+2). Hydrogen is only slightly soluble, so it is liberated as a gas.

Note the steps involved in this explanation:

1. The chemicals are identified by data criteria.
2. The reaction is described at the data level.
3. The chemicals are re-described in theoretical (Bohr atom) terms.
4. The reaction is re-described in theoretical terms.

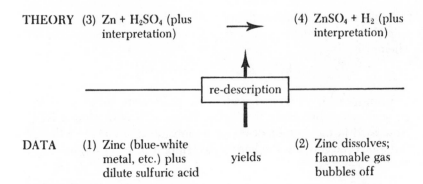

THEORY   (3) $Zn + H_2SO_4$ (plus interpretation)   ⟶   (4) $ZnSO_4 + H_2$ (plus interpretation)

re-description

DATA   (1) Zinc (blue-white metal, etc.) plus dilute sulfuric acid   yields   (2) Zinc dissolves; flammable gas bubbles off

Now, what does the theory provide that the data description did not provide? The re-description of the chemicals and the reaction allow the theorist to account for this particular chemical reaction in terms that relate it to every chemical reaction and to a set of general statements about the atomic structure of substances, their chemical valences, and how they can and do combine. The reaction in the hydrogen generator has been explained in terms of the atomic structure of the chemicals involved.

Moreover, the reaction can be "calculated" (literally, in the case of chemistry) when it has been re-described theoretically. The concepts of the theory are interlinked by a set of rules and principles: the relations of valences—the relative electromotive force of the elements. These rules and principles allow us to draw conclusions about the reactions of any substances that have been identified chemically. This is something that data can never do, no matter how much of it we gather and summarize.

The theory in this case is a well-established one, with virtually no current competitors. The re-description of the data does not need to be independently justified; the theory "has its credentials." (In the next chapter, I will pay some attention to how a given theory gets its credentials.)

The same four steps, in different order, would have allowed an experienced chemist to predict the outcome of the demonstration even if it were not so familiar an example. If the question "What would happen if I combined zinc with dilute sulfuric acid?" were asked, the steps would be as follows:

1. The chemicals are identified by data criteria.
2. The chemicals are re-described in theoretical terms.

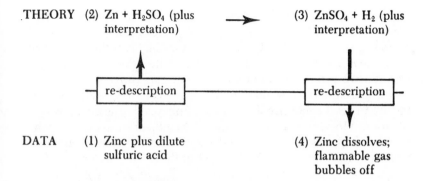

3. The reaction is calculated in theoretical terms.
4. The reaction is re-described at the data level.

Theoretical prediction involves crossing the data-theory line twice. The initial situation is re-described in terms of the relevant theory, a determination is made of the theoretical consequences of the initial situation thus described, and there must be a second re-description back to the level of data.

The fourth step is, in general, the difficult one. It is one thing to re-describe the whole demonstration in theoretical terms and account for the known outcome, as in the explanation pattern; it is quite another thing to say with certainty just how a given theoretically described state is going to manifest itself in a particular case. Predictions of this sort typically include a *ceteris paribus* disclaimer (all other things being equal) to allow for the possible things that can go wrong and alter the expected result. In this case, assuming a controlled laboratory situation, an experienced chemist would be likely to state his prediction with sufficient care to cover intervening contingencies.

The *control* in a laboratory situation is important and at the same time elusive. The aim of experiment is to isolate just the items we are interested in and eliminate all other items (variables or variable conditions) that might influence the outcome. Chemists assume "standard temperature and (atmospheric) pressure" in a laboratory experiment. If they deviate from this, they specify as accurately as possible the temperature and pressure under which a given experiment is carried out. Any chemist distinguishes between a well-run lab and a sloppily run one, and this is precisely the question of how well conditions in the laboratory are controlled: Is the ventilation sufficient to assure that gases given off in one reaction won't contaminate another reaction? Control comes right down to good housekeeping in a chemistry lab: How thoroughly is the laboratory equipment *cleaned* after each use? Minute residues from past experiments can distort the results of new experiments.

Let me return one last time to the aerosol can that was tossed into the fire to get that last predictive step emphasized in an *un*controlled situation. We have already considered two theoretical explanations of what happens when a closed container is heated indefinitely. Suppose, now, that someone is about to throw the can into the fire and is wondering what might happen if he did. A straightforward appeal to temperature-pressure-volume relationships would give the following pattern. In this case, the prediction at the theoretical level

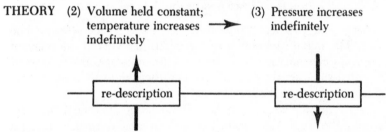

THEORY　(2) Volume held constant; temperature increases indefinitely　→　(3) Pressure increases indefinitely

re-description　re-description

DATA　(1) Closed aerosol can; heat applied indefinitely　(4) Symptoms of pressure increase

is that the pressure of the system will increase so long as the temperature is increased. What *isn't* clear is just what to expect in terms of the aerosol can. Trouble, to be sure. "Symptoms" of an indefinitely large increase in pressure. It might explode.

But there are any number of ways in which such an increase in pressure might manifest itself, depending upon the other circumstances of the case. This is not a controlled situation—and, even in the laboratory example, we had to make an allowance for circumstances other than those immediately relevant to the theory. Here we are dealing with "found" objects. The can *might* explode, or it might have a weak spot that will open and let the gas out gradually, or the plastic cap might blow out, or, if the situation is just right, the cap might melt and release the gas at the last possible instant before the can explodes.

Now, you might object that the inability to predict precisely what is going to happen is due simply to a lack of information, such as the bursting strength of the container, the melting point of the plastic cap, the precise heat of the fire, and so on. But this is no objection. It simply *is* a characteristic of uncontrolled situations that there are many such relevant but unknown variables. But we still have to predict with some degree of accuracy what will happen in such situations.

Further, it is a mistake to suppose that in even the most controlled of laboratory situations such variables can be eliminated entirely. They can only be reduced. This is underscored by the scientific community's well-known demand that any given experimental outcome be repeated in a number of instances before a new result is accepted, no matter how carefully the circumstances of the first experiment were controlled.

The difference between theoretically described states and the same states described at the data level is further underscored by the "proving" procedures that are necessary when we design a device along theoretical lines. No matter how much theoretical care we exercise in designing an electronic circuit, a rocket launcher for spacecraft, or even a lever-and-pulley gadget for a farming chore, there is the familiar business of "getting the bugs out" of the completed device. The "bugs" may be the result of mistakes or sloppy craftsmanship in some cases, and these are, of course, irrelevant to the present point. But there are "bugs" that cannot be foreseen, and they are the result of an unavoidable slippage that happens when we put our theories into practice.

It is precisely in crossing the line from theory back to data that the slippage occurs. When a known situation is re-described in theoretical terms, the connection is clear: To put the can into the heat of the fire *just is* to increase its temperature; to keep the can sealed *just is* to maintain constant volume in the system.[8] But any number of other situations might also have been identified as increasing the temperature of a system or holding the volume constant. The data-level descriptions and the theoretical descriptions don't match up in a one-to-one way. The relationship between the two is not one of straightforward translation from one vocabulary to another. The connection between theory and data is not a linguistic one, then. It is mediated through the (real or hypothetical) situation being described in much the same way that the relationships among the maps of the same territory were in the preceding chapter: each description stands in its own semantic relationship to the (non-linguistic) facts of the case, *whether or not* a direct linguistic tie can be established between the two descriptions. What the theory explains is not the descriptions couched in the data vocabulary, but the *situation* that has been thus described. The theoretical statements are not in any sense a "translation" of the data statements. The theory explains the situation by re-describing it in such a way as to connect it with other situations that are similar to it in relevant ways; and it is the theory itself that determines what counts as relevance.

The pattern of theoretical explanation just described involves a *contextual* shift in the following technical sense: The *explanandum* (data-level description) and the *explanans* (theoretical re-description) are both related semantically to the same objective situation—the same event or range of events. They make reference to that event from within different contexts of presuppositions.

The two descriptions may or may not have terms in common. If they do, the common terms will be from the relevant nomenclature ("zinc" and "sulfuric acid" in the chemical explanation). But the two descriptions are asserted about the same subject matter. They are, in linguistic terms, co-referential. Each, considered as a set of statements, is subject to judgments of truth or falsity within its own context of presuppositions relative to that subject matter.

We explain something, on this understanding of explanation, by re-describing it in a new context. *Analogy* and *modeling* also involve contextual shifts, but they are different from theoretical explanation, and we should get the distinction clear immediately. To draw an analogy is to shift from one context of description to another, but without a literal claim for the truth of the new description. To say that sound waves are like waves in water is not to say that sound waves just are waves in water; it is to invoke a similar and more familiar system (water waves) in order to clarify a less familiar one (sound waves). Analogy is what figures in the use of certain kinds of model in the sciences, as when we say that the molecules of a gas are like perfectly elastic billiard balls in order to make the properties of gas molecules more comprehensible.[9]

There seem to be no prior logical or conceptual limitations that must be placed on the forms that theories can take in order to explain by theoretical re-description. The relationships among items in the theory must be clear and straightforward in order for it to function as an explanation, and there must be at least the possibility of connecting the particular situation or kind of situation to be explained with other situations. Clearly, those elements are present in the two explanations just considered.

In the case of biological explanations, it is sometimes suggested that the theory is not susceptible to generalization because it makes reference to this planet and the history of organisms on this planet. But surely any biologist would claim that the principles of evolutionary explanation can be generalized to cover other sorts of environment. The principle of natural selection would not come under challenge; the contingent facts about this planet would simply have to be replaced with the appropriate facts about a different environment.

Any given system—a set of related items and their properties—can be re-identified as a mechanical system, as a teleological (end-directed) system, as a fluid system, and perhaps as other sorts of system as well. In the next chapter I will cite some examples of areas where such shifts have taken place; but it is worth mentioning here that electricity, for example, has been variously identified as teleological (as in "Lightning strikes because Jub-Jub is angry"), as a fluid (as in eighteenth and nineteenth-century electrical theory), as mechanical (as the movement of physical particles), and finally as a particular kind of "field." What places limits on a theory is a series of critical decisions that must be made at any given time. In the examples cited—chemistry and gases—the critical decision has been made and still stands; we are dealing with well-established theories. In the next chapter, I will discuss how such critical decisions are made and how they are altered over a period of time.

What may seem to be an even more difficult problem for this account of theoretical explanation is finding a way to decide whether or not two descriptions *are* descriptions of the same situation or the same kind of situation. The answer is essentially the same as that to the question about the forms that theories can take. Theoretical explanations are accepted or rejected after critical examination and testing. There must first be a claim that two descriptions are of the same subject matter, and the claim must be defensible in the face of careful criticism. Again, this is a matter that we must return to in more detail in the next chapter.

FURTHER READINGS

The account of explanation and of the distinction between theory and data arises from a long critical dialogue in philosophy of science, beginning with the movement called *logical positivism* in the 1930s and extending right up to the present. An excellent collection of articles by contemporary philosophers of science, which puts the contemporary views into perspective with the positivist movement, is to be found in Achinstein and Barker (eds.), *The Legacy of Logical Positivism* (1969). The literature on scientific explanation is almost entirely centered around early work by Carl G. Hempel and Paul Oppenheim, reprinted along with other relevant articles in Hempel's *Aspects of Scientific Explanation* (1965). The material which has most strongly influenced the account of explanation offered here is Richard Zaffron's "Identity, Subsumption, and Scientific Explanation" (1971).

The theory-data distinction is most strongly represented and argued along the lines presented here by Marshall Spector in "Theory and Observation" (1966). See also Karl Popper's essay "Evolution and the Tree of Knowledge," reprinted in his *Objective Knowledge* (1972); and the early chapters of his *Logic of Scientific Discovery* (1959).

The most readable and systematic source on evolutionary taxonomy is George G. Simpson's *Principles of Animal Taxonomy* (1962). See also Cain's *Animal Species and Their Evolution* (1960).

On chemistry, the material here is to be found in almost any introductory text. I have used Linus Pauling's *College Chemistry* (1957) in laying out the explanation of the hydrogen generator. On the history of chemistry, see J. R. Partington, *A Short History of Chemistry* (1965).

FIVE

The Origin of Theses

Like other species, human beings leave tracks: artifacts, ruined cities, works of art, and, unique to our species so far as we know, linguistic artifacts that record the oral traditions that preceded written language, as well as written chronicles that begin as early as 3500 B.C. and exist in increasing detail up to the present.[1] The particular line of tracks that interest us here are the tracks human beings have left in doing science.

The Critical Process at Work: Electrical Theory in 1851

I have placed a strong emphasis on the scientific community in the past two chapters. There is more to say, of course, about just which individuals constitute such a community and how they operate on new hypotheses in such a way as to accept certain ones rather than others.

In looking at the history of science, we tend to pay attention to the dramatic developments which are the direct antecedents of contemporary theories. What I propose to do here is look at the activity of the critical community of scientists as it operates on two competing hypotheses in physical theory, neither of which is accepted at present.

For the next few pages, I want to examine part of a highly respected textbook in natural philosophy (now known as physics) written in the first half of the nineteenth century by Denison

Olmsted, professor of natural philosophy and astronomy at Yale College (now Yale University). The work was well known and something over 70,000 copies were in use as a standard text before 1850. In 1851, Olmsted revised the work to incorporate new developments, "bringing the subject more fully up to the present advanced state of the science, than can probably be found in any similar work."[2] The excerpts in the following pages are from the 1851 revision.

Now, let's begin with a look at the state of the physical sciences in 1851. We can characterize it with one name: Newton. There never was before, and in all likelihood never will be again, a theory as dominant and as widely accepted as Newton's. His laws of motion, published in 1687, tied together the existing theory in a number of different areas into one neat, compact set of laws that stood unquestioned for 200 years.

Nevertheless, there were problems in the Newtonian theory that were present from the outset. The "incorporeal agencies"—light, radiant heat, electricity, magnetism, and gravity—did not fit in an easy and obvious way into the mechanical scheme that Newton laid out. It was an article of faith that they would eventually be brought into the picture in a mechanical way, and every relevant hypothesis that was given serious consideration in western science between Newton and the middle of the nineteenth century involved an attempt to explain the incorporeal agencies in terms of a mechanical system.

By 1851, there was considerable experimental data available about electricity. By far the majority of experiments had to do with what we now call static electricity, which is the most readily available but at the same time the most erratic manifestation of electrical energy.

> Although friction is the most common, and by far the most extensive means of exciting bodies, yet it is not the only means. Electricity is manifested during the *changes of state* in bodies, such as liquefaction and congelation, evaporation and condensation. Some bodies even are excited by mere *pressure*; others by the *contact* or *separation of different surfaces*. Most *chemical combinations and decompositions* are also attended by the evolution of Electricity, which manifests its presence to delicate electrometers.[3]

Although the connection between electricity and magnetism was reasonably well known, and Olmsted describes "electro-magnetic engines" (simple electric motors) and "voltaic cells" (storage batteries similar to those now used in automobiles), most of the experimental information had to do with the generation of static electricity. This was generated by rubbing glass, amber, and other materials with wool or fur, either by hand or by the use of ingenious machines in which, for example, a glass disc was rotated by means of a crank, with "rubbers" consisting of leather pads covered with silk that pressed against the cylinder as it rotated.

Figure 5-1 • A Static Electricity Machine

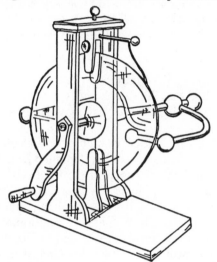

By L. Dogan, after Olmsted

Electricity from such apparatus could be stored in Leyden jars (an early form of condenser or capacitor that had been invented in 1745) and discharged all at once in spectacular spark effects. Lightning was, of course, already understood to be electricity, and due to the experiments of Benjamin Franklin (1706–1790), something of a technology had developed for using lightning rods to protect buildings.

There was also some "bad data" that influenced the explanations for electricity somewhat. Here is some bad data from a well-known source [Sir Charles Wheatstone (1802–1875), who devised the Wheatstone bridge, a type of circuit still used to measure electrical resistance]:

MOTIONS OF THE ELECTRIC FLUID

353   *The velocity of the electric fluid is exceedingly great, but its motion is not instantaneous.* Light moves at the rate of 192,000 miles in a second, but according to the experiments of a distinguished English philosopher, Mr. Wheatstone, the electric fluid, in traversing a wire connecting the outside and inside of a Leyden Jar, has a velocity of 576,000 miles per second. There is some reason for believing, however, that the velocity is different in different cases, varying with different degrees of intensity and with the kind of conductor through which it passes; for, in some recent experiments with voltaic electricity (a form of the fluid to be described hereafter) the velocity has appeared to be only 28,000 miles per second, and in others even as low as 18,000 miles per second. Still, for all ordinary distances on the surface of the earth, its motion may be accounted instantaneous.[4]

It was not until the work of James Clerk Maxwell (1831–1879) that it became established that the velocity of electricity and light were the same.[5] But this was still fifteen or twenty years away when Olmsted wrote.

The key term for the critical examination of electrical theory is to be found at the beginning of that last quotation from Olmsted: *fluid*. Since Newton himself, scientists had interpreted heat and light variously as "corpuscular" or as "hydrodynamic," the latter sometimes being interpreted in turn as involving minute particles or corpuscles. Of course, given the presuppositions of Newtonian theory, all the apparent "incorporeals" had to submit to explanation in terms of mechanical action. But *what kind* of mechanical action wasn't quite clear. In 1851, the prevailing view was that electricity, heat, and light were all "subtile fluids" and that their regular properties could be described hydrostatically. Here is Olmsted arguing hard for the plausibility of a fluid theory:

363   First, *there are some reasons derived from analogy for believing in the existence of an electric fluid.* (1.) The reasons in favor of supposing that light and heat are caused by the agency of peculiar fluids, (arguments, however, that we cannot discuss here,) which have induced a general belief, are, for the most part, equally applicable to electricity. (2.) In the present state of our knowledge, the most subtile of all fluids, indeed the most attenuated form of matter, is hydrogen gas, of which one hundred cubic inches weigh only two and a quarter grains, which is nearly fourteen times lighter than common air. But at no distant period, means had not been devised by mankind for proving the materiality of common air, nor even of identifying the existence of the other gases

which now bear so conspicuous a part in experimental philosophy. But as knowledge and experimental researches have advanced, a series of fluids still more subtile than air, have come to light, until we have reached a body nearly fourteen times lighter than air, at which, at present, the series stops. Is it probable, however, that nature stops in her processes of attenuation precisely at the point where, for want of more delicate instruments, or more refined and powerful organs of sensation, our methods of investigation, and powers of discrimination, come to their limit? An examination of the general analogies of nature will lead us to think otherwise.[6]

The "general analogies" are from the animal and vegetable kingdoms. Before the invention of the microscope, a naturalist might have supposed the smallest bit of living matter to be "the insect which is barely visible in the sunbeam," Olmsted argues; but each improvement in lens systems has revealed "successive new tribes of insects or animalcules," and there is no reason to suppose that either this succession or the succession of subtile fluids stops where our technology does.

Considerations of this nature lead us to believe that there are, in nature, fluids more subtile than hydrogen: and such being the fact, we can hardly resist the belief, that Heat, Light, and Electricity, are bodies of this class—bodies which make themselves known to us by the most palpable and energetic effects, although their own constitution is too subtile and refined for our organs to recognize, or our instruments to identify them as material.[7]

All this, of course, amounts to a defense of the prevailing scientific belief that such "subtile fluids" could exist. Now, Olmsted follows with an argument that of all the kinds of mechanical systems known, fluid systems are the most likely to explain electricity:

364 Secondly, in addition to the foregoing presumption in favor of the supposition that electricity is a peculiar fluid, *it exhibits in itself the properties of a fluid*. The rapidity of its motions, the power of being accumulated, as in the Leyden Jar, its unequal distribution over the surfaces of bodies, its power of being confined to the surfaces of bodies by the pressure of the atmosphere, its attractions and repulsions, are severally properties which we can hardly ascribe to anything else than an elastic fluid of the greatest tenuity.[8]

Two Competing Theories

Benjamin Franklin had noted in 1747 that there were two sorts of electrical charge, which he called *positive* and *negative*. An earlier text, dated 1840, and typical of its decade, is based almost entirely on Franklin's work and on the explanatory hypotheses Franklin had developed.[9] In 1851, however, there was a competing hypothesis that had originated with French scientists, and this is where the critical community really gets down to business.

Franklin had postulated that electricity was a single fluid which pervaded all bodies because of its extremely "subtile" character and was naturally in a state of equilibrium. Any given kind of body had a portion of electricity called its *natural share* and was electrically active only when it had either more or less of the electric fluid than its natural share. A body with more than its share was positively charged; a body with less was negatively charged. Electrical discharge, typically accompanied by a visible spark in the static electricity experiments that constituted the vast majority of the available data, was understood as the restoration of equilibrium. The fluid literally flowed, according to Franklin's hypothesis, from the body with more than its natural share to the body with less.

The French hypothesis, which originated before Franklin's work with Charles-François du Fay (1698–1739) and was now being pressed again by Jean-Baptiste Biot (1774–1862), postulated two distinct fluids. It was noted that the electricity that was generated by rubbing glass—vitreous electricity—exhibited different properties from that generated by rubbing amber, sealing wax, or analogous substances—resinous electricity. Vitreous electricity corresponded to Franklin's positive charge; resinous to negative charge. The two fluids were supposed to pervade all bodies, like Franklin's single fluid, and to exactly neutralize each other in unelectrified bodies. When the two fluids were separated, a body was electrified; on their reunion, the electricity was discharged, "like an acid and an alkali combining to form a neutral salt," in chemistry.[10]

> It is a remarkable fact, that nearly every electrical phenomenon may be perfectly explained in accordance with either hypothesis; nor is it agreed, that an *experimentum crucis*° has yet been found.
>
> °The "experimentum crucis" is a phrase introduced by Lord Bacon, implying a fact which can be explained on one of two opposite hypotheses, and not on the other. The figure is derived from a cross set up where two roads meet, to tell the traveller which road to take.[11]

Olmsted is hedging here, for despite the lack of general agreement, he does believe that there is a crucial experiment and that it favors the French two-fluid hypothesis. But before getting down to the crucial data, Olmsted draws on the work of proponents of both theories: Singer in favor of Franklin and Biot in favor of Du Fay. First, Singer's arguments, stated cautiously:

366   One of the latest advocates of the hypothesis of a single fluid is Mr. Singer, an able practical electrician; and the most distinguished defender of the doctrine of two fluids is M. Biot. In support of the former doctrine are offered such arguments as the following. (1.) Its greater *simplicity*. It is supposed to be more conformable to the Newtonian rule of philosophizing, "to ascribe no more causes than are just sufficient to account for the phenomena." The known frugality of Nature, in all her operations, might lead us to suppose, that she would not employ two agents to effect a given purpose, when a single agent would be competent to its production. This argument, however, cannot be applied, either where one cause is *not* sufficient to account for the phenomena, or where there is direct proof of the existence of more agents than one. (2.) The appearance of a *current*, circulating from the positive to the negative surface, analogous to the passage of air of greater density into a rarefied space. This point is much insisted on by Singer, and numerous examples are brought forward, where the progress of such a current is manifest to the senses.[12]

The examples concerning the circulation of electric currents between the inside and outside plates of Leyden jars are laid out but are dismissed as inconclusive, since they can be explained in other ways.

Now, having given Franklin's theory a fair run for the money, Olmsted brings in the data that mitigates against the one-fluid hypothesis:

The fact that *bodies negatively electrified repel each other*, (Art. 322) is a strong argument against the truth of the hypothesis under consideration. It is not difficult to conceive that a self-repellent fluid should communicate the same property to two pith balls in which it resided; but that the mere *deficiency* of the fluid should produce the same effect is incredible. This fact drove Aepinus, (a celebrated German electrician, who brought this hypothesis to the test of mathematical demonstration,) to the necessity of supposing that *unelectrified matter is self-repellent*—a supposition which is not only destitute of proof, but which is inconsistent with the general laws of nature, from which it appears that attraction and not repulsion exists mutually between all kinds of bodies.[13]

The "general laws of nature" mentioned in the last sentence are, of course, Newton's laws of gravitation. Now, one final argument to the effect that a negatively charged body cannot be understood simply to lack its natural share of the electric fluid:

In the distribution of electricity upon surfaces differing in shape and dimensions, the fluid is found to arrange itself in strict accordance with hydrostatic principles, and that too in bodies negatively as well as positively electrified. Now that the privation, or mere absence of a fluid, should exhibit such properties of a present fluid, is inconceivable.[14]

To clinch the case in favor of the two-fluid hypothesis of the French, Olmsted introduces some rather weak arguments. He clearly believes the case to be won by the argument against self-repellent matter. He argues first that when an electric spark is passed between two knobs (electrodes) that are some distance from each other, there is light at *both* the positive and the negative knobs and, if the distance is sufficient, there is no light in the middle, "where the two electricities unite." Moreover, the light given off by vitreous surfaces is different in color from that of a resinous surface, and the electrochemical effects of vitreous (positive) electricity are different from those of resinous (negative) electricity.

Now, driving the last nail home, Olmsted draws on the work of S. D. Poisson (1781-1840), whose earlier mathematical analyses of light, electricity, and magnetism were to aid Maxwell in bringing down both fluid hypotheses within a few years:

> (3.) But the most conclusive argument in favor of two fluids, is the perfect manner in which this supposition accounts for the *distribution of electricity* on bodies of different dimensions. On the hypothesis that electrical phenomena are owing to the agencies of *two fluids, both perfectly incompressible, the particles of which possess perfect mobility, and mutually repel each other, while they attract those of the opposite fluid, with forces varying in the inverse ratio of the squares of the distances.—* on this hypothesis, M. Poisson, a celebrated mathematician of France, applied the exhaustless resources of the calculus, to determine the various conditions which electricity would assume in distributing itself over spheres, spheroids, and bodies of various figures. The results at which he arrived were such as accord in a very remarkable degree with experiment, and leave little doubt that the hypothesis on which they were built is true. Nor is any supposition involved in the hypothesis itself inconsistent with established facts.[15]

Olmsted closes the argument by noting that "authority is, at the present day, almost wholly on the side of the doctrine of two fluids" and that each new discovery in Galvanism—the production of electricity by chemical, rather than electrostatic, means—and electromagnetism, won new adherents to the view.[16]

Within fifteen years after Olmsted wrote, both fluid theories had been abandoned, although the terms "vitreous" and "resinous" remained in general use until at least 1910.[17] Poisson's work had already indicated a strong connection between electricity and magnetism, and Maxwell's analysis of the equations of electromagnetic energy revealed that they were wave equations, isomorphic to those already familiar for describing the properties of light. When it was established that electricity, magnetism, and light all had the same velocity (contrary to Wheatstone's "bad data" cited earlier), it became plausible to treat all three "incorporeal agencies" as waves propagated in a medium, the ether.[18] But this was not the *only* plausible hypothesis; Maxwell himself wrote at times of "molecules of electricity," although the work of Leonhard Euler (1707-1783) in hydrodynamics, which included equations concerning "fields of force," made the wave figure more appealing.

The debate went on. If light, heat, electricity, and magnetism were to be understood as waves, then they had to be waves *in some medium*, the "incorporeal fluid" called ether. Another fluid that wasn't quite a fluid. By 1876 standard texts addressed to the student of science tended to hedge the issue altogether: "The word 'current' should not be understood to indicate the passage of a fluid, like the flow of water in a stream, but a mere transmission of the electrical force."[19] Just what the electrical force was, nobody was quite prepared to say with finality.

Even today, after relativity theory, after the abandonment of all talk of mysterious fluids, the equations describing electromagnetic activity require some hedging. Some of the properties of light are best understood as the activity of particles, others as wave functions. The compromise, which deals with wavelike "bundles" of particles, commits us to talking about particles that have a rest mass of zero— particles that aren't physical in any sense that a Newtonian would recognize and that require a suspension of common sense presuppositions about what the properties of physical objects can be.

Now, what are we to say about Olmsted and his mustering of arguments from scientists all over the world in favor of the hypothesis that electricity consists of two distinct "subtile fluids"? Is this good science or bad science? It is in the best tradition of science. Established theories and the presuppositions on which they rest are brought into the discussion of a problematic area, and hypotheses are sorted out according to how they appear against the background of well-established theory. Then the surviving hypotheses are examined critically in the light of available experimental data and sorted out once more according to how well they can account for the data that is clearly within their purview.

The Selection of Hypotheses

The debate over how many fluids one must postulate in order to explain electricity is not just a quaint episode from the relative antiquity of modern science; it is a classic example of how the community works at any given time to select one hypothesis over another. With Olmsted's careful work before us, I want now to examine the pattern that this selective process follows.

An Established Body of Theory

It is precisely because doing science is a public activity, dependent upon the use of language, that we are able to identify the theoretical beliefs of a particular place and period in history. At any given time, there is a body of theory that is accepted by the relevant scientific community, as a matter of consensus. Such established theories are the repositories of human knowledge at any given time. As statements, assented to by the scientific community at large, they rest on presuppositions that constitute the common ground of science at a given time and place and that characterize an accepted idiom.

The overall presuppositions which constituted the common ground of physical science in 1851 were those associated with Newton, as I noted earlier. The most relevant presuppositions here are the ones identified in Chapter 3 as ontological: presuppositions about what kinds of thing exist. Ontological presuppositions may be *limiting* in that they claim that there is *nothing but* certain types of entity: those physical objects composed of earth, air, fire, and water, for some of the ancients; indivisible atoms for others; and those physical processes which could be explained by Newtonian mechanics for most of western science between 1700 and 1900. Ontological presuppositions may also be *warranting*, in that they sanction assertions at varying levels of generality about identifiable entities of given kinds, such as subatomic particles in general, or quarks, or particular animal species, or "subtle fluids."

Categorial presuppositions also serve to characterize an idiom or prevailing school of thought. These are presuppositions which limit or warrant the attribution of given properties to given kinds of entity. Categorial presuppositions in general seem to be more flexible and subject to change than ontological presuppositions, perhaps because the figures of speech generated by their violation often serve as useful ways of viewing things and eventually become embedded in "literal" descriptions.

Here, then, is a basic ontological presupposition of the physical science of 1851:

> *All natural phenomena can be interpreted as the workings of mechanical systems.*

This is a limiting presupposition, and it carries with it a pledge to account for every kind of experience. It has an apparent corollary, as limiting presuppositions tend to, that is a bit ambiguous:

> *Whatever cannot be interpreted as a mechanical system is not a natural phenomenon.*

The ambiguity lies in the "cannot": do we mean "cannot *now* be accounted for" or "cannot ultimately be accounted for"? The latter is impossible to demonstrate, of course. If the presupposition is understood to refer to the present state of science, on the other hand, it carries with it the suggestion that science is at some particular moment complete. This is an arrogance to be wary of.

Clearly, such a presupposition at least affects what we take to be plausible. If it is understood as demanding that we be able to account for a given range of events now, then it is going to limit the experimental data that we are willing to admit. Only those reports which can be explained will be believed, unless they are so public and unquestionable that their plausibility is beyond doubt. Even then, we suspect trickery. Anomalous experimental results will be doubted or denied. If the presupposition is understood as demanding that we ultimately be able to account for a given range of events in a given way, then it is going to limit what we are willing to admit as an explanation for experimental data that *is* clear-cut, repeatable, and undeniable.

The notion that a scientific community constitutes a school of thought should be taken more literally than it usually is. Whether the community is comprised of a priesthood, a group of initiates, a political power, or an institutionally defined segment of the population at large, there is typically, at any place and time, a clear-cut teaching function of the scientific community that determines which individuals of the coming generation will fill the vacancies left by older individuals. What is taught and preserved is not just a body of theory, but a method and a point of view that determines what is and what is not acceptable as a theoretical explanation.

The established body of theory at a given time consists of a set of presuppositions, which are the rules for making sense within the prevailing point of view, and a set of *assumptions*, an established body of beliefs. Assumptions are primarily statements of a general nature which are taken to be true within the overall context of presuppositions. In the case of the two-fluid hypothesis, the background assumptions were the well-established theses in mechanics, hydrostatics, pneumatics, meteorology, astronomy, acoustics, and optics, to run through the rest of what Olmsted covers in his text.

If the "core" of accepted theory at a given time is a highly public matter, so, generally, are the nascent theories that are competing for acceptance and the problematic areas which afflict each theory. The problem of "incorporeal agencies" is only one such case. It was well known, for example, long before Kepler and Newton, that neither the Ptolemaic nor the Copernican accounts of planetary motion could account for the observed positions of Mars.

A problem situation always exists in an organized body of theory: either a problem with existing theory or perhaps a practical problem of interpreting data or of technique. Such problems and the hypotheses proposed as solutions to them must be understood as arising from a background of prior solutions and a more or less organized tradition of solutions.

The first element in the critical selection of hypotheses, then, is this:

Established Theory

Conjecture of a Hypothesis

A hypothesis is simply a "bright idea"—a conjecture about the possible cause of a given event, about why two sorts of event seem to be regularly connected, about what the nature of a given situation "really" might be, or about what would emerge if we looked at a situation in a particular way. A hypothesis, in other words, can be at any level of generality or at any level of theoretical abstraction from what is taken to be the data.

The conjecture we need to trace here is du Fay's hypothesis, revived by Biot after a hundred years of further gathering of experimental data and theorizing, that electricity consisted of two fluids. How it occurred to du Fay originally is anybody's guess, but it clearly arose from within the well-established structure of prevailing theory and the overall commitment to Newtonian mechanics and its presuppositions.

Franklin's one-fluid theory had been around for some time before Biot began to press the two-fluid hypothesis again. The problems with Franklin's theory were familiar, as were the experimental odds and ends that remained to be accounted for. Why does the electrical spark from amber appear to be a different color than that from glass? Perhaps it would eventually be explained on the one-fluid theory, perhaps not. How could repulsion between negatively charged bodies be explained as the mere absence of the electrical fluid? Wouldn't there have to be some agency *present* in order to cause two negatively charged bodies to repel each other?

A bright idea emerges. Suppose old du Fay was right; suppose there are *two* fluids. Let's look at it this way and see if two fluids can explain the electrical phenomena we know about *now* better than one fluid does.

A behavioral psychologist would be hard put to explain how hypotheses originate. Why it occurs to a given individual that something might be the case is as difficult to account for as why a particular genetic mutation takes place. A person might, in a given instance, be able to tell how an idea occurred to him, but for any general account we must treat hypotheses as occurring in a more or less random way against the background of an established body of beliefs.[20]

Hypothesis, in this sense, comes before deliberate experiment. Random experimental behavior occurs at the level of behavioral innovation discussed in Chapter 2, and it amounts to learning by trial and error. But the scientist doesn't run around doing random experiments, even though it happens from time to time that an unexpected experimental result in one area leads to a hypothesis in another. Hypotheses often serve as surrogates for experiments, as when we stand back from a practical problem and try to figure out what would happen if we "try it this way." Deliberate experimentation enters this pattern several steps hence. It must be preceded by hypothesis, even in the case where we are looking for statistical correlations with the aid of a computer. You can't program a computer with the instruction "find me two properties that are correlated" without first giving it the properties to choose from. "Empirical generalizations" like the gas laws, Ohm's law, or even the generalization that all redheads have freckles, must begin with conjecture—that temperature, volume, and pressure are related, for example, and that "pressure" is the appropriate way to understand that third property.

Hypotheses about the causes of particular events or kinds of events are typically influenced by the prevailing theoretical commitments about what the causes could be. But it has happened frequently that a hypothesis that flew in the face of prevailing theory proved to be successful. Such cases happen often in medicine and nutrition, where it is particularly difficult to isolate specific agencies from the overall setting of a living system and develop straightforward accounts of what is relevant to what. The development of the germ theory of disease, associated with Louis Pasteur (1822–1895), was such a case.

Theoretical hypotheses are not derived from the data in any sense, but are conjectured by the scientist in order to account for the data. Here, too, the prevailing theory will be influential, either by its strength or by its weakness. A theory without problems does not exist. There never has been and never will be such a theory. A theory that cannot be improved upon does not exist either.

In most cases, a theoretical hypothesis grows out of the recognition of a problem or puzzle in the present state of scientific theory, whether it has to do with integrating some apparently aberrant data into an established structure, or perhaps altering the structure to account for the aberrant data, or reconciling two parts of the structure itself that do not quite fit together, or even questioning on philosophical grounds some of the presuppositions on which a major portion of the structure rests.

It isn't even necessary that anyone *believe* the hypothesis at this point in the schema of selection. The conjecture that all motion could be viewed as relative was proposed first not by Einstein but by Nicolas (of) Oresme in the fourteenth century, the period most historians regard as the very beginning of modern western science. But Oresme didn't believe it himself. He was simply spinning off hypotheses as a kind of philosophical exercise to show that the rigid doctrines that had held sway until then about the solar system didn't have any unique claim to truth. This was one of several hypotheses that he proposed simultaneously in his refutation of the claim that the rotation of the heavens was an observed fact. Oresme foreshadowed Einstein in technique as well: he proposed his hypothesis as part of a thought-experiment (*Gedanken-Experiment*). Imagine an observer out among the fixed stars watching. The apparent motion would be the rotation of the earth. The observer and the stars would seem to be stationary.[21]

Here is the second element in the pattern of critical selection of hypotheses, then: Against a background of established theory and presupposition, a bright idea is conjectured.

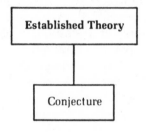

Plausibility

People get bright ideas every day, and many of them are never communicated to another person. The first challenge that a hypothesis faces is internal, in both the logical and the psychological senses. The obvious internal question has to do with the self-consistency of the hypothesis itself, although a hypothesis of any but the most gross sort can be rephrased in such a way as to be made self-consistent in the minimal logical sense.

But there is another type of internal or subjective test that a hypothesis has to pass, and that has to do with its plausibility in the broader sense outlined in Chapter 3. Is the hypothesis consistent with known relevant data? Does it fit with well-established theory? If not, does it constitute an improvement on well-established theory? Is it, in other words, the sort of statement that one would want to propose for consideration?

Olmsted lays out a hypothesis in his Section 364 that is clearly causal, clearly influenced by the prevailing theoretical commitments, totally plausible in its context, and, from the perspective of modern physics, clearly wrong. What causes static electricity to appear at the surfaces of bodies and stay there? The pressure of the atmosphere, according to Olmsted. If electricity is to be understood as a "subtile" mechanical agency, then it must, of course, be subject to the other mechanical forces present, such as atmospheric pressure.

The hypothesis we are tracing, du Fay's hypothesis that electricity must consist of two "subtile fluids," was clearly consistent with the presuppositions of Newtonian mechanics. It seemed a reasonable enough conjecture that it was worth a try. It did not in any way go to the very core of established physical theory.

The decision that a hypothesis is plausible in this sense amounts merely to a judgment that it is worth a try, that it is likely to stand up to criticism and experiment. A hypothesis that seems, on mulling over it, to be implausible or inconsistent is likely to be dropped. It may, like Oresme's hypothesis about relative motion, be picked up again hundreds of years later, but this, like the recurrence of random experimental behavior that is not reinforced, is largely a matter of chance. In the figure below [O] represents Oresme's relative motion hypothesis and [D–B] represents the Du Fay–Biot two fluid hypothesis.

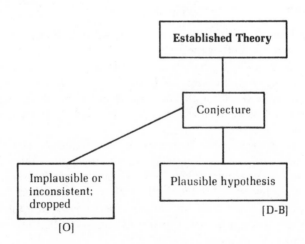

The Critical Challenge

A hypothesis has to be put in touch with reality. We do this by determining whether or not the hypothesis enables us to describe reality in a systematic way. Typically, but not necessarily, this can happen in an armchair. The autobiographical details of the innovator aren't really relevant, although they can sometimes be fascinating. But we mustn't suppose that we can have a tight description here of exactly what "going public" with a hypothesis consists of. In the medieval university, highly structured debates went on constantly, and hypotheses were put forward as a matter of scholarly duty. At other times, the process has been less formal. The custom of presenting a hypothesis at a professional meeting still persists, and this might or might not be the first objectification of a given bright idea.

The first critical examination of the revived two-fluid hypothesis in the light of accumulated evidence might have been carried out in private by Biot before he mentioned it to his closest colleague. On the other hand, he might have blurted out his bright idea impetuously, as soon as it seemed reasonable to him, and then spent a pleasant hour or two with some colleagues over a bottle of good French wine, running through all the known data about electricity they could think of. The question, however addressed, is whether or not the two-fluid hypothesis provides a way of re-describing the well-known demonstrations with pith balls and glass rods, static generators, sealing wax and cat's fur that would tie the results together and at the same time meet some of the difficulties that afflicted the prevailing one-fluid theory.

Nine times out of ten (one supposes), the relatively private criticism that a new hypothesis receives will kill it off or modify it considerably. "Sure, it works for the experiments we do in the east wing of the lab, but you can't possibly explain what the west wing people did last week that way." That's enough to send the waiter for more wine and either change the subject or start modifying your hypothesis.

Visions of the Sorbonne aside, the critical examination of a hypothesis is a matter of dialogue, whether carried out over a comfortable glass of wine or smoking typewriters. No matter how formal or informal the "going public" is, there is an ever-widening circle of critics who have a go at a new hypothesis. Whether the critical examination begins with the innovator himself, his dinner companions, his seminar, his most trusted colleagues, or a press conference is of no consequence. The critical community goes to work, and its best work often happens in conversation rather than in print.°

No matter how sweeping the implications of a new hypothesis are, this critical step is the important one—where a hypothesis becomes objective and where it becomes scientific. A new hypothesis might challenge the presuppositions that underlie a major portion of prevailing theory. It may, like Du Fay's, remain within the prevailing presuppositions and challenge the main thesis of an established theory. It may challenge a minor point of existing theory, or, finally, it may compete for a place that isn't firmly held by any contemporary theory. Such is the case at present with hypotheses in many of the "soft"—social—sciences, and it is also the case within the physical sciences to a lesser extent.

Even the decision about just what a new hypothesis is relevant to is a critical decision. Surely, any two theories that are relevant to the same domain of phenomena should be comparable, to see if either is superior in accounting for the relevant empirical data—that is, the data that we have a standard way of describing already. But just what the domain of a theory or a new hypothesis is, exactly what empirical data we should expect it to re-describe in such a way as to make comprehensible, is a matter for argument. Olmsted demanded that a hypothesis about the nature of electricity be able to account for data on static electricity, galvanic electricity, and voltaic (chemical) actions—but not magnetism or bioelectrical events in the animal nervous system and surely not light. Twenty years later, the demands on such a hypothesis were significantly different.

Franklin himself had been dead for sixty years when Olmsted wrote. There were implications of his theory that had never occurred to him, that couldn't be associated with any particular living individual in 1851. The quarrel that Olmsted was adjudicating was

---

°One of the pleasant things about Olmsted's work is that it preserves some of the critical give and take that is often lost to us. The same is true of Clifford M. Will's article, "Gravitation Theory," to be discussed shortly. Critical dialogues can be followed in some professional journals in specific fields. There are some interesting running quarrels that go on from time to time in *Systematic Zoology*, for example. Others can be followed in the interdisciplinary journal *Science*, published by the American Association for the Advancement of Science, and major issues and discoveries in a number of different fields can be followed in *Scientific American*, *Nature*, *Natural History*, and other periodicals.

not between Franklin and Du Fay or Biot; it was between two segments of the scientific community, one of which still believed Franklin's theory, the other of which now believed Du Fay's. The general cautiousness and conservatism of the scientific community in accepting a new hypothesis is of enormous importance here. The more radical the proposal, the more cautious members of the community are likely to be in accepting it, and the longer it is going to take before anything like a consensus is reached.

Beyond the scientific community, whose boundaries tend to be extremely blurred at times, there is the community at large to be dealt with. The new way of looking at reality might or might not gain adherents easily among nonscientists. Galileo had problems with which we are all familiar; so did Darwin, and so does he still. Open criticism, even open publication, of scientific hypotheses has been subverted at various times in history by church or state (in very local and very wide areas). The pressure to serve the ends of extrascientific institutions is sometimes great. This topic bears further discussion in more detail when we have other examples before us.

The acceptance of a theory can be risky. In some cases, operating on a shaky theory can imperil life, and it is not inconceivable that a given theoretical hypothesis might guide action that could imperil the whole species. There is a chilling story concerning the explosion of the first atomic bomb: just before it exploded, a scientist was asked whether or not the chain reaction might go on indefinitely and literally blow up the world. The answer: "It might." Exploding the bomb *was* the crucial experiment—the least possible crucial experiment at that time—which would determine how good some of the hypotheses were that the physicists believed.[22]

Following Karl Popper, I have called the way new theoretical developments gain acceptance the critical method. A hypothesis is proposed as a way to account for available data; it is subject to scrutiny in the light of known data and in the light of the prevailing theory in other areas. Further data suggested by the hypothesis is tested, and where possible a "crucial" experiment is designed which will give clear reasons for preferring either it or the existing, competing theory. Crucial experiments are hard to come by in the clear-cut sense that we sometimes like to think they exist. Sometimes an attempt at a crucial experiment tests only the flexibility of two competing theories to account for new data. Any Franklinian worth his salt could have explained repulsion in one way or another.

Surely, Franklin would have given it a try if he had still been around.

There is a famous example, cited by Toulmin and Goodfield, where a crucial experiment, or one which seemed to be crucial, took a twist and pointed theory in a backward direction for a short time. The competition was between what we now know as *oxygen* and *phlogiston* as the element to explain combustion. Did burning objects give something up or take something on when they burned? Surely they seemed to be lighter after they were burned in some cases (think of burned wood). Something was given off in combustion, it seemed. Call it phlogiston, and let's see what else we can find out about it.

Or, as other chemists had good reason to believe, was there something *taken on* in combustion? Careful laboratory measurements of the weight of some very pure chemicals seemed to indicate that this was the case. The weight of metallic mercury increased when it was burned to produce mercuric "calx" and decreased when it was reduced by further heating back to metallic mercury.

The experiment consisted of heating some mercuric "calx" (oxide) in an inverted jar that was filled with hydrogen (as we now call it) and sealed with water. Hydrogen was identified as "flammable air" and was taken to be that separable portion of the air in which phlogiston was most likely to be found. If the phlogiston theorists were correct, the water that sealed the jar would be sucked upward, as the mercury took back its phlogiston out of the confined quantity of flammable air in the glass jar. If the oxygen theorists were correct, the water should be pushed downward, as the oxygen was driven off the mercury calx and added to the flammable air in the jar. The outcome: The water level in the jar was *higher* at the end of the experiment. A crucial, but very temporary, victory for phlogiston.*

When it is brought under critical challenge, a hypothesis takes on objective significance. The hypothesis that stands up to internal challenges is brought out into "public space" for criticism and evaluation by the scientific community at large, including the criticism of experimentation. The deductive consequences of the hypothesis are explored, spelled out mathematically if this is possi-

---

*It is perverse, I know, to have mentioned Olmsted and Du Fay and to have kept the oxygen and phlogiston theorists anonymous. For the explanation, the theorists, and the fuller description, see Toulmin and Goodfield, *The Architecture of Matter*, pp. 226–227.

ble and appropriate, and subjected to test. The reasoning patterns involved in the critical examination of scientific hypotheses are typically longer and more complicated than those involved in the ordinary life situations discussed in the preceding chapters, but the principles are the same.

The objective significance of hypotheses depends on intersubjective testing of their consequences. A hypothesis is objective, then, and subject to judgments about its plausibility and ultimately its truth or falsity, in much the same way that the descriptive statements are that we discussed in Chapter 3. Hypotheses stand in their own relations to the facts, and need not rest on any particular mode of description of the facts. They may, indeed, demand that experimental data be re-interpreted or re-examined.

A hypothesis can be assessed for truth or falsity only if the context of presuppositions in which it occurs is understood; and this is rather like understanding the "legend" of a map. One can imagine the skeptics of 1851 jokingly asking how much it would cost to buy a quart of the electric fluid. Franklin's and Du Fay's hypotheses would have seemed a bit silly outside the context of presuppositions that prevailed in the physical sciences of the day.

As a new hypothesis is brought out into public space, the first steps are taken to evaluate it. Its ability to account for known data must be established first, before any question can arise concerning its displacing other hypotheses. Can the hypothesis provide informative theoretical re-descriptions of known experimental results that are clearly within its purview? Does it provide a way of looking at experimental results that can consistently be applied over the relevant range of the theory?

Du Fay's hypothesis clearly could. Each of the experiments which Olmsted describes can be re-described in terms of the action of the vitreous and resinous fluids. Moreover, Du Fay's hypothesis gave a better accounting for the repulsion of negatively charged bodies than Franklin's did.

In the process of challenge and further experimentation as appropriate, a plausible theoretical hypothesis may be unable to provide an adequate re-description of clear-cut data or of problematic or questionable data, or it may provide explanatory re-descriptions of all known data, as the du Fay–Biot hypothesis did in 1851.

A *causal* hypothesis, at the data level, must stand up to the same challenges but without the re-description step. Does smoking cause cancer? How well does the hypothesis that it does stack up with available data? The questions are the same.

Here, schematically, are the possible situations at this fourth step in the critical process.

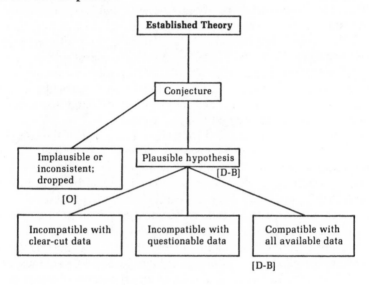

A Thesis Emerges

A hypothesis becomes a thesis when members of the scientific community become convinced that it is true. If a hypothesis fails to gain adherents, it simply drops out from the mainstream of scientific belief. Think of Oresme's suggestion in 1370 that all motion is relative. His hypotheses were written down, and we know about them, but they failed to gain any adherents. Oresme himself did not believe that all motion was relative. The "intellectual ecology" of western thought, if I may be permitted the figure, was all wrong for this hypothesis in 1370. But when its time came, in a period of scientific crisis at the beginning of the twentieth century, this basic tenet of relativity theory took hold with a vengeance.

When a thesis gains adherents and becomes established as at least one respectable hypothesis (perhaps among several competing ones), what happens next depends on the strength of the new thesis in comparison to its established competitors, and on how important it is to the prevailing established theory.

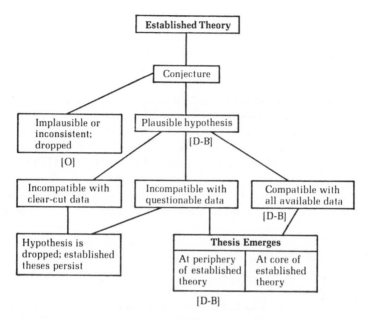

Where Does the New Thesis Fit?

Du Fay's two-fluid hypothesis did not strike to the heart of prevailing physical theory. It challenged nothing in the overall Newtonian presuppositions concerning the eventual development of mechanical explanations for all natural phenomena. It did not challenge any assumptions, any well-established theses in mechanics, hydrostatics, pneumatics, meteorology, astronomy, acoustics, or optics.

What it did was challenge a respected but problematic hypothesis at the periphery of the prevailing physical theory. It clearly stood up as well as Franklin's hypothesis, and the outcome of the critical dialogue recorded by Olmsted went one of two ways: One of Du Fay's or Franklin's theses would win out, and the other would be dropped; or, more likely, the two would coexist at the periphery of the main body of theory for a time, literally increasing the diversity of scientific belief until further theses were developed which either made the choice even more clear-cut than it was in Olmsted's argument, or wiped out both of them in a more radical change.

Maxwell's later hypothesis that light, electricity, and magnetism were all energy that moved in waves came closer to challenging the core of prevailing theory. (The hypothesis is represented as [M] in the diagrams.) Waves that were not in any obvious sense waves *in some medium* seemed to be a new form of energy, not easily accounted for in Newtonian terms. It was well-known that light "waves," if that was what they were, would pass through the most complete vacuum that could be created in a laboratory situation. But the Newtonian presuppositions were strong enough and well-established enough that the wave theory was quickly absorbed into the main body of theory by the revival and slight modification of the notion of *ether* (or *aether*): a "subtile fluid" that pervaded *all* space and served as the medium for the transmission of wave energy. Newton had speculated at some length about such a fluid but hadn't really required it for any of his theses. Now it was needed; it was consistent with existing theory in other areas, and it served to explain Maxwell's wave theses and to absorb them into the existing body of theory.[23]

It is fair to say, I think, that the ensuing attempts to find independent evidence for the existence of the ether are what led to the crisis in physical theory that culminated in the special and general theories of relativity.

The progress of Einstein's hypotheses can be traced down the right-hand side of the opposite figure, represented as [E]. Einstein's theses challenged Newton's, which were right at the core of established theory, however problematic the theory had become by the end of the nineteenth century. This entailed an across-the-board challenge of both presuppositions *and* assumptions, because the ontological and categorial presuppositions of nineteenth century science were dominated by the Newtonian conceptions. Men like Olmsted took them to be "the general laws of nature." Every assumption of physical science, the well-established theories in all the fields Olmsted covered and more, rested on the Newtonian presuppositions.

The most dramatic possible outcome of the critical process occurs when a major theory is displaced. The existence of problematic areas in an established theory will not bring that theory down. The only thing that can displace a theory is a stronger theory. On the relatively small scale of the debate between the one-fluid and the two-fluid theories of electricity, we can see some of the considerations involved in such displacement. Two fluids explained the repulsion of negatively charged bodies in a satisfactory way; one fluid did not.

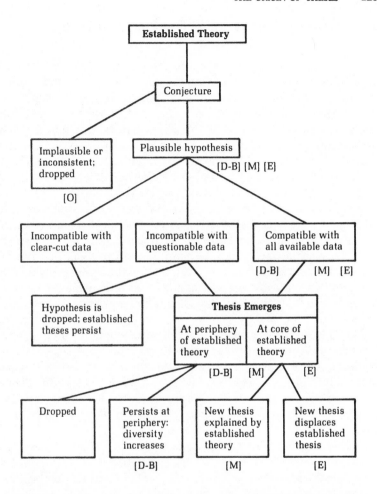

Einstein's theory absorbed Newton's. It explained everything that Newton's did and more. It accommodated the areas where Newtonian theory had difficulty, and moreover it explained why Newton's theory had worked as well as it had. It allowed the accurate prediction of planetary motions that Newton's theory had and, in addition, made accurate predictions about the orbit of Mercury which Newton's theory could not make. But such culmination is never permanent in science. The equilibrium is always problematic, and it always will be.[24]

The Critical Process Still at Work: Gravitation Theory in 1977

The present state of equilibrium in the physical sciences is not as easy to describe as was the state in 1851. The special and general theories of relativity, quantum theory, and the recent developments in elementary particle theory have absorbed Newton's work, in the sense that they explain all of Newtonian physics within a larger physical theory, with presuppositions different from Newton's. But we can't "freeze frame" on the present the way we could freeze the state of physical theory accepted by the scientific community at large as of the time Olmsted set down his pen in 1851.

We are inclined to think that we didn't know until quite recently what light, magnetism, electricity, and radiant heat are, and that we are pretty well along in knowing what gravity is. But if knowing what they are means having a theory to explain them that has no gaps in it, we still don't know. Even if the critical process allows us to claim that we know something at a given time, what we know is no more firmly established than Olmsted's knowledge that electricity was fluid in character.

The "incorporeal agencies" are still problematic to some extent, because of the wave-particle compromises mentioned earlier; but all of them except gravity have been brought into relation with each other. That "one last detail"—gravity—promises to be a difficult one to account for, and accounting for it is likely to shake up the existing theory as much as Maxwell's work shook up the fluid theories of electricity.

A recent article by Clifford M. Will, a physicist at Stanford University, describes the current competing hypotheses about gravity in a way that is strongly reminiscent of Olmsted's description of the competing hypotheses about electricity:

> Today many physicists firmly believe that the general theory of relativity, which explains gravity as a curvature or warping of space and time, is the correct theory of gravitation. They praise its beauty, simplicity and agreement with observation and experiment. Other workers, however, are openly dubious of general relativity and suggest that alternative theories provide a better description of gravity. Some have devised theories of their own and expound on *their* beauty, elegance and agreement with observation and experiment.[25]

Will sets out to delineate which of the many competing hypotheses about gravity are "viable" (he uses this term where I use "compatible" in the schema). His criteria for a viable theory of gravity can be paraphrased as follows:

1. It must agree with observation and experiment; that is, it must account for all well-established observational data and new observations that are within its critically defined scope. (Will calls this criterion "completeness," a term which I prefer to avoid here because of its uses in logic.)
2. The theory must be consistent; it must not predict incompatible results. One theory is rejected as not viable, for example, because it predicts one result when light is treated as a particle and a quite different result when light is treated as a wave.[26]
3. It must be compatible in its nongravitational aspects with the special theory of relativity, which is well-confirmed in *its* nongravitational aspects by particle theory.
4. It must be compatible with the observed properties of matter and the "gross," or average, motion of the sun and the planets as described by Newton. Presumably, it would explain and predict the detailed structure of the planetary orbits in such a way as to account for the "gross" motions predictable by Newton's laws of gravitation.

These criteria seem quite stringent, and one would expect the choices to be narrowed down to just a few. But Will describes *eight* competing theories that are viable by his criteria, of which Einstein's general theory of relativity is only one. And each of the remaining seven represent *groups* of competing theses that differ within the grouping on relatively minor points. Moreover, these eight groups represent only those viable theories in which gravitation is treated as curvature in space and time. There are yet other groups of theories (called "nonmetric") which treat gravity otherwise and which may or may not conflict with a problematic set of experiments carried out between 1889 and 1908 in Hungary.[27]

Will argues that only two of the eight otherwise viable groups of theories stand up to the results of the most recent solar experiments, but he treats this matter with proper caution—remember Wheatstone's "recent experiments" in Olmsted's account. The two are Einstein's theory of general relativity and the "scalar-tensor" theories associated with R. H. Dickie and Carl H. Brans. He suggests, finally, that one of these or a theory not yet formulated will be what gains general acceptance in the coming decades.

Since Will wrote in 1974, there have been flurries of excitement in the scientific journals concerning the possible identification of particle tracks in bubble chambers as "gravitons"—particles of gravity. Nothing is settled about these particles, and it is a problem, as Will noted, for the coming decades.[28]

A Theory About Science Itself

As you probably realized some pages back, the sequence of steps distinguished in the critical evaluation of hypotheses comprise yet another set of stories that follow Darwin's overall plot for selective change as I analyzed it in Chapter 2. The patterns from the history of science that I have been describing as a set of related critical activities can be re-described in the vocabulary I used earlier to describe the evolutionary development of bodily and behavioral change. This bears some clarification and explanation.

I warned you early on that what I have in mind here is a theory about science. An explanatory theory, as I described it in Chapters 3 and 4, involves a step of re-identification, as when we say that the temperature of a system *just is* the mean kinetic energy of the molecules that comprise that system, or that gravity *just is* a curvature in space.

Let me make the claim as explicit as I can; I am not going to suggest that the development of knowledge systems in human populations is *like* the selection of bodily and behavioral traits or analogous to it in some schematic sense. The thesis is stronger than that; I am suggesting that the systematic development of human knowledge is *part* of the overall process of natural selection and adaptation that accounts for the bodily and behavioral traits of all species.

The schema laid out in Figure 5-2 is not a third phase of evolution, to be added to the phase of endosomatic evolution which governs bodily change and the phase of exosomatic evolution which governs behavioral change. The critical communal process of scientific inquiry and theorizing *just is* exosomatic evolution. Exploring a hypothesis, asserting it, testing it, criticizing it, believing it, and acting on it *just are* systematically related behaviors. As carried out by organized *populations* of human beings, they evolve as other patterns of behavior do—under the environmental pressure of natural selection. They are yet one more instance of the way in which a particular species, over a period of many generations and

Figure 5-2 • The Selection of Hypotheses

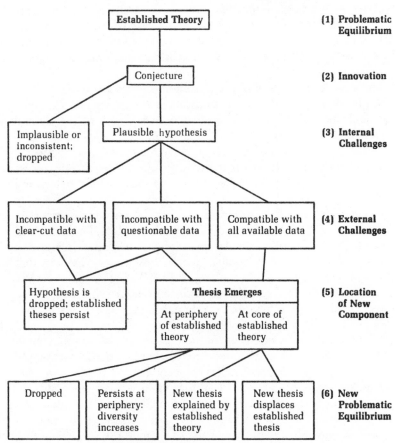

under the selective pressure of the facts of life, has evolved a set of techniques for dealing with the facts of life.[29]

Theories do not evolve, of course, except in the same derivative sense that spider webs, beaver dams, and bird's nests do. What evolve are the physical and behavioral characteristics of living beings. Theories, as recorded sets of statements, do not in any literal sense have a life of their own. We may speak figuratively of the evolution of theory. We may be fascinated for an afternoon at a museum by a display of the various steps in the "evolution" of the steam engine, the can opener, or indoor plumbing. But what literally evolve as part of the mainstream process of natural selection are patterns of behavior. The environmental facts that exerted pressure on human beings to organize in social groups, to develop languages, and to codify information in particular ways have also led to a set of identifiable activities for developing group solutions to common problems.

Some more theoretical re-identifications before we go on: Having a bright idea, conjecturing a hypothesis, *just is* experimental behavior, and the critical selection of hypotheses *just is* the natural selection of behavior patterns. The neural and sensory characteristics that are common to all human beings *just are* species-specific physical traits, and the formal presuppositions that I claim to be rooted in those physical traits *just are* species-specific behavioral traits.

Beyond the formal presuppositions, there is enormous diversity in the presuppositions that characterize the conceptual, linguistic, and scientific behavior of particular populations of human beings. To take up what is by now a familiar theme, it is this very diversity that makes a wide range of adaptation possible. Take this seriously: We *are* the most flexible species on earth, no matter what people say about cockroaches. They are flexible physiologically. We are flexible conceptually and technologically.

What human beings have in common physically allows us a much wider variety of effective intellectual behavior than most people think (or than most people think is respectable). We tend to view the intellectual style of our own tradition as in some sense uniquely correct. This is the attitude I described in Chapter 1 as immature.

The activities associated with individual scientific traditions are subject to enormous variation on the surface, in the same way that other physical and behavioral characteristics of living beings are, and for the same reasons. We are a global species. The geographical isolation of populations of human beings that led to the development of distinct languages and cultures has also created for each population a distinct set of problems that must be addressed and a distinct set of solutions to those problems. The order in which problems are addressed in a given population has also influenced the divergence. Once a given set of solutions to problems is established, it forms the background for, and influences the outcome of, each subsequent hypothesis that arises within a scientific tradition.

How can we speak of there being *one science* amid all this

diversity? There are two ways. The one I have concentrated on in this chapter is this: scientific activity is one particular kind of communal activity which occurs in all human populations. I have described it in the schema of Figure 5-2. It is part of the core of the established repertoire of group behavior of the human species, and it was established long ago, probably not long after the development of language itself. The development of group solutions to common problems is made more effective by the systematic criticism of hypotheses and beliefs. The same selective pressures of the facts of human life which were discussed in Chapter 2 have favored the development of a particular repertoire of group activity in human populations; I have called this repertoire scientific activity. It includes codifying and communicating beliefs and hypotheses and subjecting them to disciplined critical judgment in a systematic way. Many individual intellects are brought to bear on hypotheses which affect the community at large.

In that discipline of judgment resides the conservatism of a scientific community. Any hypothesis has to be subjected to the critical challenge of a community of tough-minded specialists before it is generally accepted and integrated into the overall body of beliefs which guides the actions, plans, and expectations of individuals within the overall population. Deliberate alterations of habitual actions are accompanied by critical alterations of belief.

It is precisely the systematic criticism and testing of hypotheses against a background of established beliefs and experiences that distinguishes scientific activity from unscientific activity. Just how the community of tough-minded specialists is constituted varies from one time and place to another, as I will illustrate in the following chapter.

What makes a hypothesis scientific? The answer to this is not to be found in the structure of the hypothesis itself but in the question whether or not it can be dealt with scientifically. It surely does not depend on any distinction between theoretical hypotheses and data-level hypotheses. The hypotheses mentioned previously were all relatively theoretical—the two hypotheses about electricity, Maxwell's hypotheses, Einstein's, and the theories of gravity considered by Will. But the development of trial and error technology follows the schema of exosomatic development as well. Against a background of habitual techniques for, say, constructing an ox cart, a bright idea occurs to someone for a possible improvement. Is it plausible? If so, try it out, either by discussing it with others or by subjecting it to test. Will it work? Will it work better than the familiar technique? The range of questions and challenges is the same as those for a theoretical hypothesis and, as noted earlier, for generalizations.

Is there a particular subject matter that we can isolate as scientific? I think not. There is no area of inquiry that cannot be addressed scientifically. The question must be about the way in which the subject matter is addressed, not about the subject matter itself. It is a mistake, then, to ask whether or not astrology, for example, is scientific. Astrology can be done in a scientific way, if its hypotheses are addressed critically in the way I have described in this chapter. Too often, they are not.

Are there unscientific hypotheses, then? Only those which cannot be criticized and tested.

Finally, what do we demand of a theory? What must a theory do? Olmsted and Will give us the answers here, in their critical work. A theory must fit in; it must be consistent with the established body of theory in other areas. But more important, a theory must provide a way of re-describing the data within its scope in such a way that the data is made intelligible and related to a larger body of beliefs.

I submit that the theory I have proposed here to explain human scientific activity is a plausible one, in the sense described above, and that it explains human scientific activity by re-describing the components of that activity as a set of systematically related behaviors.

I have drawn on only a small number of examples in laying out this theory, and these are all from the familiar fields of western science. This is only the first round of critical challenges a theory like this must face.

In the following chapters I will suggest the second way in which this theory allows us to speak of one science. I will draw on the history of scientific thought in both eastern and western science to test the theory's ability to account for a wider range of data, to explain the relationships among disparate scientific traditions, and, finally, to point out ways in which these scientific traditions are converging as a result of the breakdown of the geographical barriers which have separated human populations. The thesis that distinct traditions of knowledge can converge, and are indeed converging, remains to be argued in the chapters that follow.

FURTHER READINGS

Denison Olmsted, the author of the text in natural philosophy employed in this chapter, was born on June 18, 1791. He was educated at Yale College. In 1817 he was elected to the professorship of chemistry at the University of North Carolina at Chapel Hill, and in 1821 he was appointed the state geologist and mineralogist by the state of North Carolina. He returned to Yale as professor of mathematics and natural philosophy in 1825. He devoted most of his time and attention to astronomy and was well known for some of his papers on meteor showers. His several books on natural philosophy provided him with considerable income. He died in New Haven on May 13, 1859.

The background to Olmsted's work is covered succinctly in Toulmin and Goodfield's *The Architecture of Matter* (1962), Chaps. 9 and 11. More detailed background is to be found in Sir Edmund Whittaker's *A History of the Theories of Aether and Electricity* (1960).

The way in which hypotheses arise is discussed by Ghiselin in *The Creative Process* (1955), and by Koestler in *The Act of Creation* (1969).

Material on Nicolas Oresme is to be found in Toulmin and Goodfield's *The Fabric of the Heavens* (1961), Crombie's *Medieval and Early Modern Science* (1959), and a short selection of his work is included in Ross and McLaughlin's *Renaissance Reader* (1953).

The overall theory of science presented here is strongly influenced by Karl Popper's works, *Conjectures and Refutations* (1962), *The Logic of Scientific Discovery* (1959), and *Objective Knowledge* (1972). Quite different, and opposing, points of view are to be found in Paul Feyerabend's *Against Method* (1975) and Thomas Kuhn's *Structure of Scientific Revolutions* (1962). A critical exchange between Popper and Kuhn is included in Lakatos and Musgrave, *Criticism and the Growth of Knowledge* (1970).

SIX
Holism and Pluralism as Philosophies of Nature

The great physicist Werner Heisenberg has suggested that the most fruitful developments in human thinking occur at the times when two different lines of thought, rooted in different parts of the world, come together.[1] I am going to deal with two lines of thought in this chapter that I will characterize broadly as eastern and western.

This is simplifying matters, of course. Although I have been speaking freely of western thought, western culture, and western science throughout the past chapters, we know perfectly well that there are interesting and important distinctions to be made among the conceptual styles of different language and cultural groups within western civilization, as well as in art, and, yes, science as well.

But we don't have to back off very far from those differences to speak in comprehensible terms of the traditions of culture, religion, art, and science that can be located historically in the Mediterranean area 2,000 years ago and traced through to populations that now live in all of Europe and the Americas.

Similarly, although there are important differences among the identifiable Asian schools of thought—Confucianism, Taoism, Buddhism, Hinduism, and others—there are strong enough historical and philosophical interconnections among them that we can speak sensibly of one traditional culture of the Far East.[2]

What we have to deal with are two *clines*, as I introduced the term back in Chapter 2 (p. 47): two subgroups within the overall human species which differ from each other in identifiable ways. There are other clines which can be distinguished within them, but it is sufficient for my purpose here to concentrate on the two.

Philosophies of Nature

There are several circumstances which distinguish the sort of cline I am discussing: its languages and its developed line of solutions to problems, obviously, and its political and military history as well, as I will discuss shortly. But to give a fair characterization of the divergent lines of thought between the two major clines I am interested in here, we will have to characterize the philosophical view of nature which is peculiar to each.

Olmsted's use of the term "Natural Philosophy" as the title of his physics text was not idiosyncratic. The term was in common use among scientists until the beginning of this century. Olmsted opened his first chapter thus: "Natural Philosophy is the science which treats of the Laws of the material world." He continues, "The term Law, as here used, signifies the mode in which the powers of nature act. Laws aim at determining things with numerical precision, or of assigning the exact proportion in which effects take place." A philosophy of nature—a way of looking at nature—pinning down the laws of the material world, is what western thinkers have been after from the time of the ancient Greeks. At times, as early as 500 B.C. (Heraclitus) people have argued about whether or not it was possible to find scientific truth at all. At other times—for example, right after Newton's work—people have argued that scientific truth had been found once and for all.

A question which runs through a good deal of western thought about other cultures is this: Why is there no science elsewhere? Why, specifically, is there no science in the East?

I am not going to try to convince you that the occult mysteries of the East are scientific after all. What I am going to argue is that a good deal of oriental thought is neither occult nor mysterious and, beyond this, that in some areas it may be more scientific than the characteristic western thought on the same subject matter.

The first step is to look at some of the important presuppositions of western and eastern thought in general. What will emerge is worth foreshadowing here: the distinction between mind and body which has been characteristic of western thought implicitly since Plato in the fourth century B.C. and explicitly since Descartes in the seventeenth century A.D. just never caught on in the East.

Modern western science has been characterized by a sharp separation between areas which are the "proper subject matter of science" on the one hand and mental and spiritual matters on the other. The proper subject matter of science has been understood to include physical, publicly observable events interpreted on what is usually called a *materialistic* model. There can be no quarrel with the claim that the striking successes of western science and technology since the seventeenth century are due in large part to precisely this point of view. It seems obvious to many people that the enormous strides taken in the physical sciences indicate that there is something essentially right about physics, chemistry, biology, and related fields as they have been approached in the West for the past 300 years. But there have been striking failures as well in western science, and these are gaining attention at present. It is precisely in the areas having to do with human beings themselves that western science has had the most trouble. Beyond this, physics in the twentieth century has brought under question the very materialism which has characterized western science as distinct from other traditions.

In Chapter 1, I suggested that the confusion of ideas which confronts us today is a result of the recent rapid and detailed communication of ideas. This shouldn't suggest that the major intellectual traditions were totally out of contact with each other until very recently. There has always been some interchange of ideas among human cultures, but in the past, influences have tended to be subtle enough and slow enough that they were assimilated gradually. They were influences between two populations of the species which lived in separate environments. This is no longer the case. In the past 200 years, the many environments of human beings have come close to being one environment.

What I am suggesting about the convergence of lines of scientific thought isn't all that dramatic, then. There have been innumerable times of contact between clines, often brought about by commercial trade or by territorial conquest. Influences have been exchanged. What differentiates the present from the past in this respect is the rapidity with which influences are exchanged between clines.

In both the eastern and the western clines, systematic thought is linked to a heritage of natural philosophy. This heritage is perpetuated by means of the teaching function of a scientific community mentioned earlier. An overall philosophy of nature is gradually modified by a critical process of the same sort as that involved in the modification of the particular theories which arise *within* a natural philosophy.

We do not approach nature without questions, without hypotheses to be criticized and tested. Such hypotheses always occur against a previously established background of presuppositions, some of them so basic as to constitute the prevailing philosophy of nature and some of them incidental to specific bits of theorizing.

What begins as a hypothesis sometimes becomes so thoroughly woven into the fabric of scientific thought that it *becomes* one of the presuppositions which defines the context in which further hypotheses and theorizing take place. Moreover, such presuppositions eventually become *limiting* presuppositions and define an orthodoxy which can lead at times to ignoring or denying relevant empirical data. But limiting presuppositions have their positive effects as well: a successful line of theorizing, a well-understood context in which to frame hypotheses and theories, must be explored to its limits before it is abandoned.

The Newtonian world view did not fail to catch on because it could not immediately account for the incorporeal agencies. It became, as I noted earlier, an article of faith that the incorporeal agencies *would* be accounted for within the Newtonian presuppositions, and properly so. Until a *new* context of presuppositions arises which can successfully challenge the old, a relatively closed system of thought develops, which isolates itself from other approaches and other systems of knowledge. The failures of the system do not bring down the whole school of thought. It wasn't the ultimate failure of the Newtonian system to account for the incorporeal agencies which brought down the Newtonian orthodoxy. The orthodoxy stood until a new set of hypotheses, demanding presupposition change, stood up to the critical challenge. When the hypotheses of relativity physics zoomed down the right-hand side of the schema of evolutionary change as I have represented it (on p. 129) and replaced Newtonian mechanics, a set of presuppositions concerning the nature of the world was replaced by another. This is the Darwinian plot at its most dramatic; it is the exosomatic equivalent of Darwin's phyletic change, as described in Chapter 2.

Over the centuries, the questions, the hypotheses, and the prevailing presuppositions of scientific communities have changed. The Ionians in 600 B.C. asked about the *ingredients* of the world. Galen in 200 A.D. asked about the *function* of the items he distinguished in nature, and his questions were framed mainly within the context of a study of the human body. Between 1500 and 1600 A.D., the questions turned to the *mechanism* of the world. In our own century, the questions have changed again, in ways that cannot be characterized with quite the glibness that centuries of hindsight give us on the earlier changes. Each period of change in the basic presuppositions which characterize a philosophy of nature brings about dramatic shifts in scientific thought and in the *habits* of thought which are eventually reflected throughout an entire population.

Science cannot be viewed as developing along a straight line, as an accumulation or accretion of discoveries, any more than an evolving lineage of organisms can be viewed as an accumulation of individuals. Scientific development is not always "progress." There are many blind alleys in any evolutionary pattern. The evolution of habits of belief is, like other aspects of evolution, a matter of differentiation, specialization, and reintegration—a progression from unity through increasing diversity to more complex patterns of unity.[3] This pattern has been repeated many times in both the eastern cline and the western.

Two Clines and Their Histories

A cline has a history, and that history includes more than the broadly characterized philosophies of nature which I must concentrate on here. The political and military facts of a cline's history influence the choice of scientific problems at any given time. The major developments in nuclear fission and fusion and in space technology within the past forty years testify to this. There are times when particular problems must be addressed by a scientific community because of the press of extra-scientific events. There have always been such times. Science just is the development of group solutions to common problems, and the problems often arise for reasons that have nothing to do with science itself.

The most remote causes of the differences in habits of thought in distinct clines are easy to speculate on but difficult to be sure of. Certainly the climate of northern Europe and the needs to store energy in the forms of food and fuel for the winter had something to do with the problems that were of first and most immediate concern to groups which settled there. But, in point of fact, there are some striking parallels between eastern and western thought lurking just beneath the surface idioms, at several times in history right up to and through the fifteenth century. Without any pretense at developing an intellectual history of East and West, I am going to divide the relevant history of the two clines into four periods prior to the period we call *modern science.*

Antiquity to 600 B.C.
600 B.C. to 200 A.D.
200 A.D. to 1100 A.D.
1100 A.D. to 1700 A.D.

What I propose to do is set the stage for some observations about modern science by paying attention to the main lines of thought which preceded it and gave form to the context of established beliefs into which modern science had to be integrated.

Antiquity

Many writers on the history of science tend to view the sixth century B.C. as the beginning of scientific thought in both the East and the West. Before that, it is suggested, was a time of primitive technological developments and superstition. But this period of time, prior to the periods for which we have reasonably detailed written records of scientific and philosophical thought, forms the setting against which the dramatic developments of the sixth century B.C. took place.

The technology which arose before the earliest available written records bespeaks a good deal of activity which must be called scientific. The problem solving of organized groups of human beings does not happen by fortuitous accident but by a critical process of trial and testing that follows the overall plot outlined in the previous chapter. Within the context of an established set of solutions to problems, new hypotheses arise, come under critical evaluation and experiment, and, if they prove successful, eventually have their effect on the overall context itself.

As early as 4000 B.C., copper was being smelted from ores in Iran and Israel by means of fairly complicated devices for heating the ore in the presence of other minerals and extracting the metal. Some of the natural copper ores contained compounds of arsenic, and these, on smelting, yielded a hard alloy similar to what we now call bronze. But these very ores proved dangerous to deal with (as we explain it now, because of the arsenic which made the smelted metal harder than pure copper). We have no records of how the problem situation was conceived, or of how the ancients explained the fact that the metal extracted from some ores was harder than that extracted from others. But we do know that a solution to the problem of dealing with the dangerous ores was arrived at long before the arsenic-laden copper ores ran out. The solution involved using the less dangerous ores which ordinarily produced softer copper, but mixing tin with the copper extracted from them to form the more familiar bronze alloy.

The tin had to be imported to the eastern Mediterranean region, and, for reasons which aren't quite clear, the supply of tin was cut off quite suddenly at about 1200 B.C. (The reasons seem to be related to political and military events, but the records are ambiguous.)

Another problem, then: how to get a harder metal? Iron ores had been used incidentally in the smelting of copper, and these now proved useful. The higher temperatures required for extracting iron from its ores demanded more extensive use of charcoal to fire the smelting furnaces, and it was noticed that iron from some parts of the furnace was harder and more durable than that from others. There is no way of knowing how it was explained at the time, but as we would now explain it, *carbon* from the charcoal had become mixed with the iron in minute amounts, producing a kind of steel. Track the lines of reasoning, of hypothesis and testing, of explanations we can only guess at, and the sudden appearance of carbon steel in the common artifacts of the eastern Mediterranean at about 1200 B.C. becomes comprehensible. By 900 B.C. carbon steel artifacts were commonplace in Iran.[4]

This is problem solving on a community basis, well before written records are available to tell us how people thought or how the processes were understood. We have only the results.

But there is another area where artifacts and a few fragmentary written records help to give us a sense of how people conceived nature prior to 600 B.C. The movements of the stars have been recorded and observed for a much longer time than surviving written records indicate. Structures such as Stonehenge in England, Teotihuacan in Mexico, Angkor Wat in Indochina, and ancient Babylonian and Egyptian structures all indicate that they were built in such a way as to line up with regular appearances of the sun, the moon, and the planets. The regular appearance and reappearance of particular lights in the sky was predicted on a year-to-year basis with great accuracy. Some Babylonian written records from about 625

B.C. indicate that by then even the looping (retrograde) motion of the planetary bodies as well as eclipses of the moon were predictable with great accuracy.

In most civilizations, the lights in the sky were associated with deities. Is this superstition? The relationship between various conceptions of deity and conceptions of reality will recur several times in the remaining chapters. What I want to argue here is that the nature gods associated with celestial events represent the beginnings of theoretical science and form the background against which the dramatic events of the sixth century B.C. must be understood.

It is easy enough to understand the many attempts to identify the lights in the night sky and explain their motions. But why, one wonders, do so many early explanations invoke deities? Why identify the lights in the sky with *that* kind of entity in particular?

The first thing we have to note is that the study of astronomy as we know it was not distinguishable in early writings from the study of climate. Meteorology had to do with everything "up there." In the main, events "up there" that were of immediate concern seemed to follow no regular pattern. The day-to-day weather, times of drought, flood, disease, or famine, seemed unpredictable and capricious—as capricious as a willful person. The analogy is a compelling one. As regularities were gradually recognized, the natural deities, as personalities, figured in the explanation of climatic and celestial events. Much later, commenting on this period, Aristotle observed that a person has a nature, a pattern of behavior that makes him more or less predictable with increasing acquaintance; and if any personalities are true to their own nature, then the gods certainly must be. (We also speak of certain acquired skills or habits as becoming "second nature," as contrasted with the "first," more basic nature of a person.)

So, when the Greek myths say that Demeter (the wheat harvest) springs from the union of Zeus (the Sun, the ruling celestial god) and Gaia (the Earth), there is more than a figure of speech present; the gods have a theoretical status. Understanding their various personalities allows explanation of events and possibly thereby the prediction and manipulation of events. Some of the personalities are relatively reliable—like the Pleiades, the daughters of Atlas, who sail high in the sky to mark the time of harvest and low in the sky to mark the time for plowing and sowing. Others, who rule the sea and the weather, are ill-tempered and unpredictable. Understanding nature meant identifying and understanding the superhuman personalities who ruled nature, deciphering their signals, and trying to propitiate them.

According to Hesiod, an early recorder of such matters, even the day-to-day matters of living—of food, sleep, and medicine—were understood in terms of superhuman personalities; applying a medicinal herb was not connected directly to any causal efficacy on bodily states but was understood rather as a means of assuring the presence of the relevant deity. (Notice that this is how such matters were explained, not how it was discovered that certain herbs had healing properties.)

Once it is taken for granted that an acceptable explanation must be in personalistic terms, a context exists in which explanations can be offered. Theological explanations were perhaps the first theoretical explanations: the pattern of re-description and re-identification noted earlier as a signal of theoretical explanation is present when a particular group of celestial lights is re-identified as the Pleiades, the daughters of Atlas, and we are told what we might expect of them and how what they do will affect us.

It is partly because of the early nature gods that religion and science became linked as early as they did. To manipulate nature meant getting the relevant deities to cooperate. Particular operations or techniques were often *explained*, like the early medicines, as invocations of the gods. This immediately sets apart a group of technical experts in the persons of priests and interpreters of the deities' personalities.

What survive of the period prior to the sixth century B.C. are the artifacts of a technology, fragments of contemporary written records of celestial events and harvests, and records of religious beliefs and religiously connected codes of conduct, many of them constituting the rules of civilized life, but as many others constituting rules for personal hygiene, diet, and even techniques for ensuring that the gods would be present to guarantee that the pottery was fired successfully or that the crop was sowed and harvested at the proper time. (For example, the dietary laws in the Old Testament Book of Leviticus constitute excellent nutritional techniques for the geographical area, even in the light of modern theories of nutrition.)

Forecasting was particularly important as human beings formed permanent settlements and civilizations. Successful crops depended in large measure on successful understanding of the gods' intentions.

The detailed astronomical records of the Babylonians, the Chinese, and others provided guides to planting and harvesting and to anticipating tides.[5]

This, then, seems to be the key to the period of nature gods and to the prevailing natural philosophies of the period before 600 B.C.: the forces of nature were *personalities*. To understand nature was to understand those personalities, to ensure their cooperation wherever possible by using techniques which proved to be effective, and to predict their activities where they seemed to be beyond cajoling by careful observation and recording of their habits.

Is this science? It certainly isn't twentieth century science, at least not on the surface of it. But beneath the surface lurks a familiar pattern. Let me step aside from what we know about early natural philosophy for a moment and construct a religious myth and a fictional religious explanation. So as to insult no particular ancient deity, we'll use an invented one: Jub-jub, the ill-tempered god of the storms—unpredictable, perfectly vile at times, striking about with his lightning, which does mankind no good at all and can only be understood as a manifestation of Jub-jub's great wrath. Many civilizations sacrificed food and precious materials, even human lives, to propitiate their storm gods. But suppose, just suppose, that one of these civilizations used thick strips of copper decoration at the corners of its important buildings, running vertically from the highest points of the tallest buildings down to their very foundations. And suppose that someone noticed that Jub-jub's great flashing wrath seemed to be manifest all about these buildings but that somehow the buildings themselves and their contents weren't damaged, weren't torn to shreds as other buildings were.

Here we have a problem, framed against a background of established beliefs. Why? What is it about the copper-cornered buildings that neutralizes the deity's wrath? This calls for explanation by the specialists in understanding the forces of nature, and in this case the specialists would constitute a priesthood. While the priests pondered the problem and criticized each other's suggestions within the overall context of beliefs about the nature of the deity, you can be sure that there would be a severe drain on the local supply of copper strips. When the explanation did come, it would have to be consistent with the rest of what was believed about Jub-jub, and the best explanation of all would be one which fortuitously explained some of Jub-jub's other characteristics in terms of the tranquilizing or distracting properties of copper as well. But an electric *fluid*? *Two* electric fluids? As out of place then as they are now. The orthodox pattern of explanation demanded an understanding of a personality, not a mechanical agency.[6]

The Classical Period

Beginning in the sixth century B.C., there are records available which give us a more detailed sense of how nature was understood and of how systematic attempts to explain natural events were undertaken. In the West, this was the century of the Ionian nature philosophers— Thales, Anaximander, and Anaximenes—of Xenophanes and Heraclitus, and of Pythagoras. In the East, it was the century of Confucius, of Gautama Buddha, and of Lao Tzu.

The Ionians wanted to know how the universe was composed, what its basic ingredients were. Behind the constantly changing flow of events, they postulated that unchanging principles must exist, and they made serious attempts to understand those principles without associating them with particular personalities.

Why did this particular turn of thought happen at this particular time? The Mediterranean world was relatively stable at this time politically; travel and commerce became widespread. Greece was where the trade routes converged, where people from a variety of backgrounds came together. In that very contact of people from different backgrounds lay the beginning of a new set of hypotheses about nature. Xenophanes in particular had traveled widely enough himself and had met enough of the various foreigners who passed through Greece that he became aware of the differing traditions of nature gods in Greece, Persia, Ethiopia, Egypt, and other regions; and he saw past the particularity of the many local nature gods to the general principles that they personified.[7]

Xenophanes was not alone in his denunciation of the anthropomorphic traditions. It was a general time of looking past the capricious, variously personified forces of nature for a more comprehensible characterization of what the forces were. Again, coinciding in East and West, an interest in mathematics and geometry developed, and it is in this period that we find the first well-documented attempts at a grand synthesis of human knowledge of the world, which has been compared to the Newtonian synthesis 2,200 years later.[8]

Figure 6-1 • Major Figures and Events
During the Period 600 B.C.-200 A.D.

WEST   EAST

BC

Thales [640-546]   Chou Dynasty
Anaximander [611-549]   — 600 —
                                        ——— Lao Tzu [604-531]

                                        ——— Buddha [563-483]
Pythagoras [580-500]   ——— Confucius [551-479]
Anaximenes [556-480]
Heraclitus [535-475]   — 500 — Chuang Tzu [c. 490] *Tao te Ching*

Socrates [470-389]   ———
Democritus [460-370]   — 400 —
Plato [429-344]   ———

Aristotle [384-322]   ——— Nei Ching [Chi Ni]
Alexander the
Great [356-323]   ———
Academy at Alexandria   — 300 —
Euclid [330-260]*   ——— Mohists
Aristarchus [310-230]*   Experimental
Archimedes [287-212]*   Science, Geometry Ch'in Empire [221]
                                                        Burning of the Books [213]
                               — 200 —   Han Empire [202 B.C.-9 A.D.]

Rise of Rome Hipparchus [c. 190-120]* ——— Alchemy—
                                            "Practical Chemistry"

                               — 100 —

Alexandria under
Rome   ——— Sunspots recorded
Birth of Jesus   *BC*
                               AD   Hsin Dynasty
                                                         Later Han Dynasty
Britain under Rome
Hero [fl. 62]*   ———

                               — 100 —
Ptolemy [85-165]*   ———
                                        Wei Po-yang [fl. 142]
Galen [130-200]*   ———

                               — 200 —

                               — 300 —

*Members of Academy at Alexandria

Pythagoras (580–500 B.C.) was a geometer, and the school of thought which bears his name constituted an attempt to tie together medicine, religion, astronomy, and music on a geometrical model. The overriding presupposition of the Pythagoreans was that all natural phenomena could be accounted for geometrically, that the spatial and mathematical relationships they explored in the abstract could provide a systematic way of dealing with the world which would tie together all human experience. The Pythagoreans were a major influence in Greek thought for almost 300 years, until the time of Aristotle (384–322 B.C.). At the end of the period of the Pythagoreans' strongest influence, Aristotle, in *de Caelo*, criticized them severely for trying to force natural phenomena into the framework of their geometrical theories, for doubting or denying the validity of data which didn't fit the overall world picture they had created. This illustrates another familiar theme throughout the history of natural philosophy: the prevailing presuppositions of Pythagoreanism constituted limiting presuppositions, which defined an orthodoxy that eventually had to break down.

But Pythagoreanism, despite its long influence, was not without competitors. Heraclitus, who was roughly contemporary with Pythagoras himself, represented quite another point of view. It is sensory information on which our beliefs about the world rest, said Heraclitus, and we can never claim any knowledge beyond that. We live in a world of constantly changing sensory impressions. Everything is in a state of flux, and the world is a jumble of disconnected facts. It was Heraclitus whom I mentioned earlier as despairing of any well-formulated theoretical science.

Heraclitus introduced into Greek thought two important themes which have striking parallels in the East: the doctrine of constant change and the view that change is to be understood as the dynamic interaction of opposite forces. Cold things become hot; moist things become dry. Things are always in a state of transition from one state to another opposite state. This is the nature of things, not imposed by some outside agency but implicit in the character of the world. No geometrical laws can capture this; it must be understood by each individual as the way of nature.

Just a generation before Heraclitus, but half a world away, was Lao Tzu, whose teachings about the constantly changing flux of events and the dynamic interplay of opposites are so strikingly similar to those of Heraclitus that in English translation it is sometimes difficult to distinguish the two men's sayings.[9]

Lao Tzu is generally regarded as the founder of *Taoism*, a school of thought which pervaded oriental thought for almost 2,000 years and is in constant revival. The union and interaction of opposites *yang* and *yin*, the demand that nature be understood on intuitive grounds by contemplating its constant dynamic interplay of opposite characteristics, and its *way* (Tao)—understood as a natural order which defied strict codification, to be followed by each individual without a well-defined set of general laws—all have their counterparts in the views associated with Heraclitus. Both philosophers are spoken of as mystics in the sense that both held that each individual must reach his own understanding of nature's way—the *Tao* for Lao Tzu, the *Logos* for Heraclitus. Nature is to be followed, not led.

Was there a common influence on the two schools of thought? There isn't any clear external evidence that there was. More likely, the conditions in the two parts of the world developed along parallel lines as trade and travel increased, and the striking and fascinating parallels in their work are to be explained by common changes in the environment.

Here I have to give only slight mention to very important philosophical figures of both East and West. Plato, whose teacher was Socrates and whose pupil was Aristotle, developed a philosophy of nature which grew literally out of the dialogues of the opposing views of his time. From Heraclitus came compelling arguments that the world was ever changing and in constant flux. From Parmenides and *his* pupil Zeno came strong, self-consciously *logical* arguments that change was unthinkable and illusory; that reality had to be understood as permanent and unchanging. From the Pythagoreans came the tidy and unifying notions of a geometrical structure at the base of all reality. The dialogue of Plato which deserves particular mention here is the *Timaeus*, because, for a long period of western history, this was the only remnant of the Platonic heritage which was available for study.

The theme of Plato's synthesis of the opposing philosophies of nature is present in the *Timaeus*: the world has a rational and logical structure that men can discover by means of their intelligence. Plato's inspiration was the geometry of the Pythagoreans. An account of the natural world would be intellectually satisfying only if it took the form of a mathematical system. Concepts were to be understood in the abstract, as *forms*, as pure and as free of individual variation as the geometrical abstractions of squares, triangles, cubes, and octahedrons. Here was eternal stability. The shifting, changing material world, the functioning parts of the bodies of living beings, had to be understood through a prior understanding of the forms themselves. But even here the geometrical model was called in. Plato took the atoms of the basic elements to have the characteristic shapes of the regular three-dimensional geometrical solids.*

The final major influence from Greek philosophy on the crucial periods of western science is Aristotle. Aristotle's own interests were in marine biology, and his philosophy of nature departs from Plato's along precisely the lines that one would expect from a biologist. Living beings cannot be accounted for as mathematical or geometrical abstractions, he argued. One must speak of their nature and their function, of the living principle that directs them in their actions. Nature, for Aristotle, was alive, an organized totality of entities, each developing according to its *own* natural characteristics toward its own destiny. What Aristotle sought to discover, and what he viewed as the proper thing for science to ask about, was the natural order of things. Aristotle picked up from his predecessor Empedocles the notion that there are four basic elements of which everything is composed—earth, air, fire, and water—but he added a fifth, the *pneuma*, or essence, in which structure and activity—the organizing principles that tie the other elements together into things of specific kinds—were to be found. In animate begins the pneuma carries the *psyche*, which develops in such a way as to direct their activities.

The corpus of Aristotle's scientific work is staggering. He addressed himself to logic, physics, biology, and astronomy, working out a systematic theory of nature consistently and in great detail and, at the same time, setting down a philosophy of nature that pervaded western thought, in constantly reinterpreted and revived forms, right up to the year 1600. (Like Plato, Aristotle wrote extensively on other topics as well, maintaining consistently the approach he took to scientific matters.)

Aristotle was both a philosopher of nature and a theoretical scientist. He argued for basic presuppositions about nature and human experience of nature and then went on to develop, *within*

---

*Plato is best remembered not for his writings on natural science but for the dialogues in which he puts together a coherent view on equality and justice and on politics, aesthetics, and ethics. These matters are, of course, outside our direct concern here.[10]

these presuppositions, detailed scientific theories. His explanations of motion bear particular attention here, because they were very influential centuries later. In inquiring into the natural order of things, Aristotle developed an account of physical movement based on what he took the constituents of matter to be. The two relevant properties of matter are *heaviness* and *lightness*, and objects are understood as being relatively heavy or light depending upon how they are constituted out of the four inanimate elements—earth, air, fire, and water. The pure element earth has absolute heaviness and no lightness; the pure element fire just the opposite. Air and water are lighter than earth, but heavier than fire.

The weight of complex objects is a relative matter. A piece of wood that weighs, say, a pound is heavier than a one-ounce piece of lead if they are compared in the elemental medium air; but compare them in the elemental medium water, and the wood is lighter than the lead. The relative heaviness and lightness of two objects then depends not just on the relationship between the objects themselves but on how each of them relates to the medium in which their heaviness is compared.[11]

In general, heavy objects fall to earth because they are composed primarily of earth, and that is their natural place. (Newton didn't explain *why* gravity acts on objects, either, remember.) The natural motion of objects is either toward the center of things (earth) or away from it (fire). The downward movement of heavy objects makes right angles with the earth's surface, and it is therefore directed toward the center; and the center of everything, of course, is the center of the earth itself.

But the motion of terrestrial objects does not go on indefinitely. Objects seek their natural place in relation to the center of things. Fire, being lighter than air, rises to its proper level above the air but does not continue rising to infinity. *Horizontal* motion on the surface of the earth is always the result of the interaction of bodies. Objects which are running the course of life on earth are constantly changing according to their proper structure or activity (the pneuma). They move horizontally only when acted on by other bodies (effectively, on a horizontal surface, when you stop pushing or pulling an object it stops moving). They move vertically to reach their natural place relative to the center of things. But the motion of celestial bodies is another matter entirely. Unlike terrestrial objects, the heavenly bodies are permanent and unchanging; they are in unceasing, constant, regular movement, with no beginning and no end, perfectly symmetrical, and inevitably—the whole body of reasoning demanded it—*perfectly circular.*[12]

During Aristotle's lifetime, Philip of Macedonia gained control of Greece. Philip's son, who was later to be known as Alexander the Great, studied with Aristotle. The conquests of Philip and Alexander spread the Aristotelian philosophy of nature throughout the western civilized world and on into Asia. Aristotle's presuppositions defined the context in which scientific reasoning was to take place. The natural order of things was known, thanks to Aristotle's theories, and only relatively minor debates about them took place at the great center of learning established at Alexandria, in northern Egypt. The "animal spirits" in things—the pneuma and psyche of Aristotle—were understood. Except for sporadic quarrels with the teachings of the master over details of specific points of theory, everything worth knowing was known. The action turned from theoretical science to applied science as the great military upheavals of Alexander's empire and, right on its heels, the Roman Empire, demanded.[13]

The academy, library, and museum at Alexandria were established and protected by the dynasty of Egyptian rulers founded by Ptolemy, one of Alexander's generals, after the death of Alexander in 323 B.C. Aristotle's theories were preserved and consolidated here, and practical researches, measurements, and observations were carried out, while the rest of the world was engaged in the wars and conquests of the Roman Empire, until 400 A.D. Those associated with the Alexandrian academy include the great mathematician and geometer Euclid (330-260 B.C.); the astronomer Aristarchus of Samos (310-230 B.C.); the mathematicians and engineers Archimedes (287-212 B.C.) and Hero (first century A.D.); Hipparchus (160-125 B.C.), whose accurate astronomical observations laid the groundwork for the later theoretical developments of Claudius Ptolemaeus [Ptolemy (85-165 A.D.), no relation to the dynasty of rulers]; and the great physician Galen (130-200 A.D.).

Alexandria came under Roman rule in 50 B.C., and increasingly the scientists were pressed into the service of the empire. Ptolemy was assigned to draw up maps of the world, Galen to organize the systems of medicine and healing, Hero to design mechanisms for military and colonial use. Under Roman influence, the interest turned away from theoretical matters and toward the development of practical devices to serve the needs of the growing and constantly troubled empire.[14]

Natural philosophy in the Orient followed lines that are surprisingly parallel at times to the lines of thought in the West. Chuang Tzu, continuing in the tradition of Lao Tzu, is credited with writing down the book which bears his name late in the fifth century B.C.[15] It is a commentary and development of the *Tao Te Ching*, associated with Lao Tzu himself. Chuang Tzu elaborates the theme of the interplay of the opposites yang and yin as the principle which guides nature and which must be understood if nature is to be understood.

At the same time, Buddhism and Confucianism continued to interact with Taoism, and the mutual influences extended beyond China to India as well. Confucianism, with its respect for everyday common sense and reason, encouraged the "investigation of things" as the source of knowledge, and this was blended with the Taoist influences in a continuing synthesis and interaction.

Roughly contemporary with Aristotle, late in the fourth century B.C., a dialogue is recorded between the natural philosopher Chi Ni (Chi Yen) and his king. In a discussion of the influence of natural events on human lives, he says the following: The vital force (Ch'hi) of the elements metal, wood, water, fire, and earth alternately influence the events of the world, but all these changes are natural fluctuations in the fundamental regularity of yin and yang. The cycles of nature and the changes we experience are to be understood as part of the natural order of things. The yin and the yang within individual things gives them their fixed composition and motions with regard to other things in the web of nature's relationships.[16]

This is the organic model of Aristotle, rendered in the oriental idiom. It continued as a central part of oriental thought in both China and (as a similar view in yet another idiom) in India, and it was never truly abandoned, as Aristotle's fundamentally organic understanding of nature ultimately was in the West.

In the third century B.C., at about the time of the establishment of the Alexandrian academy, there was a sudden turn in China to applied science, again because of the press of military events. China was in upheaval, and rival kings fought back and forth across the country. A group of natural philosophers called the Mohists undertook what is regarded as the nearest thing to modern empirical science that China ever produced.[17] They wrote on inductive and deductive logic; they described research methods based on the principles of agreement and difference for the investigation of natural regularities; they sought to apply logical principles to zoological classification; they developed optics, astronomy, and mechanics; and they attempted to define such notions as work, energy, time, and space. But the movement lasted less than a full century.

In 221 B.C. the Ch'in Empire was established, and for the first time most of China was under a single government. Shih Huang-ti, the first emperor, tried to unify the country by building a network of roads and canals and by uniting the defensive walls of northern China into the Great Wall, which delineated the boundary between the empire and the nomadic barbarians of the north.

In 213 B.C. the emperor attempted to reduce the many styles of writing within his empire to a single style and to unify the knowledge of the scattered traditions. His measures were extreme and are written of with some bitterness by Chinese scholars today. He burned all the books, except for those in his personal library, and reputedly slaughtered more than 450 scholars whose work did not meet with his approval.[18]

As in the West, theoretical science came to a halt, and the full attention of scholars was focused on the needs of unifying an empire. Through the following centuries of political and military upheaval, Taoism remained the dominant spirit of scientific thought. Atomistic, mechanical approaches to understanding matter were tried and rejected several times, but the Taoistic understanding of nature on an organic model, characterized by the principles of the vital force Ch'hi (strongly analogous to the pneuma of Aristotle), and the understanding of the interplay of the opposites yin and yang dominated Chinese thought.*

In another of the historical parallels, there was an outstanding Chinese physician, Wei Po-Yang, who worked during the period around 140 A.D., simultaneous with Galen in Alexandria. He understood the body and its medicines in terms of the five oriental elements and the principles of yin and yang. Like Galen, he studied physiology and observed the pulse and blood types.

The Middle Ages

The period between 400 A.D. and 1000 A.D. is regarded by many scholars as the "dark age" of western science.[20] During this long

*The concern with language itself and with the distinction between things and the properties of things, both of which are present as problems in Greek philosophy, do not receive much attention in oriental thought. Some writers attribute this to the pictographic character of Chinese writing, in which the connection between individual symbols and concepts is more direct than in the phonetic alphabets developed in the West.[19]

period, scientific thought was constrained by political and religious upheaval. It was for western scientific thought a time of preservation, not of development.

The major area of intellectual concern was theological. The questions became questions about the nature of god. Theology was the "queen of the sciences," and the Christian theology of Europe was under attack from the Muslim world.

Little was known in Europe of the scientific thought of the Greeks. Plato's *Timaeus* was known, in Latin translation, and Aristotle's logical works were available in a translation made by Boethius (480–524 A.D.), but the rest of the Greek and Alexandrian work was unavailable.

The Greek heritage eventually moved both eastward and westward. From the eighth to the eleventh centuries, the action was all in the Arab world. All of Plato's and Aristotle's works were available there in Syriac translations from the sixth century onward. When the great Muslim conquest of the seventh and eighth centuries swept through the area surrounding the western, southern, and eastern shores of the Mediterranean, extending from Spain to India, it absorbed and carried with it the intellectual traditions of Plato and Aristotle, as well as Galen, Ptolemy, and Hippocrates. At the same time, it absorbed and blended in traditions from India and from the Far East as well.[21]

The Muslim alchemists explored the nature of matter along lines that had been established at Alexandria. Medical knowledge from Galen and from the Indians and Persians as well was collected and systematized. Mathematics took an important turn in the development of algebra during the ninth century. Astronomy and astrology flourished, as studies of the influence of celestial bodies on human events. New ways of projecting maps were developed.

In Europe proper, with little of the ancient heritage available, theoretical science was pretty much at a standstill. Religious and political matters had center stage, and the critical dialogue about the nature of the world was silenced for a time. The Greek heritage was to serve as the set of background beliefs for the development of modern science in Europe when the time came, but most of the Greek works came to Europe in the thirteenth and fourteenth centuries in Latin translations from *Arabic*, along with Arabic alterations, commentaries, and further developments of the Alexandrian and Greek lines of thought.

Technology and crafts continued to develop in both East and

West. Roger Bacon credits the Chinese with the following long list of "contributions" to western technology: printing, gunpowder, the compass, the seismograph, the wheelbarrow, mechanical clockwork, the casting of iron, stirrups and the horsecollar, and segmental arch bridges. It was the Chinese as well who developed ways of preparing grain foods for later use. The pasta which we associate with Italian cooking was an oriental development.

Through the middle of this period, Europe was loosely organized politically. What tied it together was the Christian religious heritage which was now under constant siege from the Muslim world. What intellectual activity there was took place under the auspices of a religious institution which was defending itself and extending its influence northward.

The chemistry of the period stayed within the overall Aristotelian view that matter was composed of four basic elements united by a pneuma. Chemists—alchemists, as we would call them now—in both the Arab and European worlds sought to discover means for converting one substance into another by recombining the basic elements. The famous attempts to turn base metals to gold seemed perfectly reasonable things to try, on the Aristotelian understanding of the composition of matter.

The Revival of Critical Debate

During the twelfth century, the Muslims allowed European scholars to travel to Syria and translate the entire body of available Greek works into Latin. The writings of Plato and Aristotle were available in Europe from this time on, along with writings about the earlier philosophers, the Pythagoreans and Heraclitus, and the Alexandrian scientists, including the astronomer Ptolemy.

From these versions, Thomas Aquinas (1225-1274) elaborated and extended Aristotle's physics and, in general, revived for European scholars the Aristotelian world view. At this same time, Europe began to come alive intellectually. Universities were founded at Bologna, Paris, Oxford, Cambridge, and Salerno.

From the time of the Holy Roman Empire, the Roman church dominated all intellectual activity. It was in the monasteries that ancient documents had been preserved and copied through the worst centuries of political upheaval. Aquinas' task, then, was not just to be true to Aristotle in explaining the newly available physics;

Figure 6-2 • Major Figures and Events
During the Period 1100-1800 A.D.

WEST	AD	EAST
First Crusade [1096]	−1100−	
		Chu Hsi [1130-1200]
Muslims allow translation of Plato and Aristotle	−1200−	Genghis Kahn captures Peking [1215], then Russia, Poland, Hungary; Mongols dominate Asia
Roger Bacon [1214-1292] Thomas Aquinas [1225-1274]		
	−1300−	
		Beginning of Ming Dynasty [1368]
Beginning of Renaissance Nicolas Oresme [d. 1382]	−1400−	
Gutenberg: printing in Europe		Chu Hsi revival
Nicolas Copernicus [1473-1543] Luther, Calvin: The Reformation	−1500−	Wang Shou-jen [1472-1529] Taoist revival [1522-1566]
Tycho Brahe [1546-1601] Galileo [1564-1642] Johannes Kepler [1571-1630] Rene Descartes [1596-1650]	−1600−	Western missionaries, trade, interference
Copernicus' work banned [1616] Isaac Newton [1642-1727]		End of Ming Dynasty [1644] Ch'ing (Manchu) Dynasty [1644-1912]
	−1700−	
Benjamin Franklin [1706-1790] Immanuel Kant [1724-1804]		
	−1800−	

he had to show that the presuppositions of Aristotelian theoretical science were consistent with the Christian theology of the time.

This is the setting for the next great series of theoretical developments in western science. Some historians view the scholastic period, which is associated with the revival of the Aristotelian world view, as the beginning of "three centuries of sterility and stagnation."[22] But the new Aristotelian orthodoxy is better seen, I think, as the reopening of the critical debate in theoretical science. The presuppositions and theories of Aristotle had dominated western scientific activity for fifteen hundred years. But since the destruction of the library at Alexandria, the domination had been *in absentia*; the works had not even been available for perusal, let alone systematic criticism. Aquinas spelled them out once more, and in the newly established universities the critical debate began.

I have already noted in Chapter 5 the argument of Nicholas of Oresme concerning the question of the motion of the earth. Oresme began the work of changing the Aristotelian view of a fixed and stationary earth at the center of all things—not by denying the doctrine, but by showing that it had no unique claim to truth, that neither rational argument nor observation led inevitably to that view of the earth.

A century later, Nicolas Copernicus (1473-1543), a Polish monk, studied in Renaissance Italy and was exposed to the newly translated and revived Pythagorean view that the sun was the center of planetary motion. Copernicus was later asked by the Pope to help with calendar reform, and he culminated a lifetime of study by writing, at the age of seventy, an attempt at a coherent and simplified account of planetary motion. He tried to show that the sun, the moon, the earth, and the known planets could be viewed as a coherent system in a simpler way than Ptolemy's geocentric account, and the Aristotelian presuppositions on which it rested, would allow. Copernicus postulated that the sun, not the earth, was the center of all things, and he showed that this produced a simpler account of the observed motions of the planets than the former view did. This was presupposition change, but this was not modern astronomy—not yet. Copernicus and all the following theoreticians until Johannes Kepler (1571-1630) held to other Aristotelian presuppositions, in particular to the presupposition that the natural motion of all celestial bodies must be in perfect circles.

In the Orient, throughout this same period, there was constant political and military activity. There were territorial wars among the Chinese and between the Chinese and the Muslims. The Mongols swept back and forth across the country. No great centers of systematic learning were established until the period corresponding to the late Middle Ages in Europe. Taoism remained the prevailing line of scientific thought, and the influences of Indian thought were felt as well. The conception of the world as a living, evolutionary process was present in writings available from the eighth century.

In the eleventh century, at just the time when the full corpus of Greek writings became available in the West, there was a turn toward empirical study and away from the preservation and interpretation of the ancient writings. According to the twelfth century philosopher Chu Hsi, things must be set apart from the knowledge of things. The spirituality inherent in matter was to be understood as simply the presence of the human mind in the world. To understand the true nature of things, you must look at things from the point of view of things; if you look at things from your own point of view, you will see only your own feelings.[23] Chu Hsi's great dictum was "Observe!" Despite a brief later revival, the demand for neutral observation, for divorcing the human mind from the objects it studies, never caught on.

Once more, there was extensive military action. Genghis Khan captured Peking in 1215, and within the next twenty-five years he had extended his conquests through Russia and into Poland and Hungary.

In 1368, China was recaptured from the Mongols, and the great Ming Dynasty began. The artistic and cultural renaissance of China which began in this period was terminated only by the incursions from the West in the seventeenth century.

The neo-Confucian philosopher Wang Shou-jen (1472–1529), a contemporary of Nicolas Copernicus in the West, is regarded as the major natural philosopher of the Ming Dynasty, and his philosophy of nature dominated oriental scientific thought until the eighteenth century. He returned to the Confucian common sense influences and blended them with the Taoist philosophy of nature. Wang's first attempts to arrive at a philosophy of nature consisted of an attempt to revive the teachings of Chu Hsi. Chu Hsi had said "Observe!," and Wang spent a great deal of time (some say as much as three years) studying a bamboo grove before he came to the dramatic realization that a sharp distinction between the mind and the objects it attends to would not serve any purpose. Nature must be studied critically; particular objects must be understood as the objects of consciousness. This is not to say that things exist only in the mind but that knowledge of things can exist only as a function of human awareness. The bamboo grove by itself could tell him nothing. His own will had to be engaged. He had to decide what to pay attention to if he was to come to know anything about the world.

So long as we conceive things as being independently constituted as what they are outside ourselves, said Wang, the physical world and mind are separated, and their unity is inconceivable. Knowledge that can be systematized and communicated is trans-subjective; nature must be studied critically, and the forms of thought of the human mind—those which are common to all human minds—must be brought to bear on questions about nature. Nature does not formulate either the questions or the answers; knowing subjects do. The intuitive knowledge of our own nature that is common to all human beings is inseparable from our knowledge of the world. There is no meaningful difference between knowledge which comes from "inside" and knowledge which comes from "outside," so long as it is examined in a context of trans-subjective criticism.[24]

In Europe, as the work of the earlier Greek thinkers became available and popular among scholars, rival lines of thought began to arise. Plato's works and those of the Pythagoreans became influential; Copernicus and others began to challenge the Aristotelian world view as interpreted by the scholastics. The humanistic tone of the Renaissance began to catch hold throughout Europe. The astronomical debates had begun, but toward the end of the sixteenth century, the study of material things on the earth's surface was just reawakening.

At this same time, the political and intellectual authority of the Roman church came under attack from other quarters. Martin Luther and John Calvin broke off from the established church for reasons which had nothing to do with science and initiated Christian religions in Europe which were not subject to the authority of Rome. Henry VIII in England did the same. The equilibrium shifted as the Roman church began to defend itself once again and to pull back from its encouragement of the relatively new kinds of debate which had begun about traditional doctrines. The Inquisition, representing a conservative movement within the church, sought to enforce the traditional Aristotelian view of the world as well as allegiance to Rome.

Nonetheless, beginning around 1600, the hold of the Aristotelian scholastics on scientific theory was broken. The background of established theory and presuppositions was no longer unified. The solid intellectual authority of the church deteriorated, and with it the neat certainty that was present in the works of Aquinas. This was to be the century of Harvey, Gilbert, Kepler, Galileo, Pascal, Descartes, Leibniz, Huyghens, and Newton.

Orthodoxy tried to reassert itself in the face of both theological attack from the Protestants in northern Europe and scientific criticism throughout Europe. In 1616, the writings of Copernicus, and by implication those of Galileo and Kepler, concerning planetary motion were placed on the Roman church's "Index" of forbidden books.

But it was too late. The work of these men and of Tycho Brahe, with whom Kepler had studied, was too compelling. A major turning point in theoretical science had already been reached. A scientific community—a critical community, independent of any but the most pragmatic political and religious allegiances—was established in Europe. The setting was ripe for new hypotheses to develop without religious or political constraints, subject to criticism and testing within the scientific community itself. When publication of new hypotheses was prevented by religious institutions, the action moved to northern Europe—to Holland and England and Scandinavia—where the publication and discussion of new scientific ideas could not be prevented by the conservative elements based in Rome.

There are several western candidates for the title "father of modern science," all of whom have some claim to the title: Copernicus, Galileo, Descartes, Newton. Setting aside for the moment the scientific achievements of each, let me concentrate on René Descartes (1596–1650) in particular as the pivotal figure in the West, for the simple reason that he, more than the other candidates, wrote at some length about what became the dominant philosophy of nature in western science from 1600 to 1900.

Descartes was a geometer, and he is generally regarded as the originator of analytic geometry. He was committed to the hypothesis that the structure and operation of the universe was mathematical in character—not merely in the sense that mathematical relations were to be discovered in the study of physical systems, but in a stronger sense, reminiscent of the physical conceptions of Plato. Descartes was convinced that mathematics was the *sole* base on which a theory of nature could be founded, and that all properties of material objects could be reduced to geometrical and mathematical properties. To him, mass and form were mathematical dimensions rather than physical concepts.*

But, although his physical theories are of only incidental interest, Descartes' writings on methodology and the nature of knowledge both reflected and influenced the thought of his period. The doctrine of Descartes' that was most influential in the rise of modern western science was his sharp, unequivocal distinction between the mind (soul substance) and the material world (physical substance), which included the human body.

It is always dangerous to speculate on a scientist's or a philosopher's motivations, but the political and religious climate of Descartes' time must have had some influence on his thoughts concerning the proper methods of arriving at scientific knowledge. Galileo's confrontation with Pope Urban and the Inquisition came to its climax in 1632. Descartes at this time was preparing a treatise in natural science for publication (*Le Monde*). Upon hearing of the condemnation of Galileo's work, Descartes suppressed the treatise.[25]

At the beginning and end of each of his major works, Descartes appeals to the church for acceptance of his philosophical views on nature. The feature of these views that is of primary importance here is the mind-matter distinction. Descartes argued that there are two parallel worlds, one of mind and the other of matter, each of which can be studied without reference to the other. The "I," for Descartes, was the thinking self, the soul, and he argued on several grounds that it is distinct from the body, influencing it through willing the body to action and being influenced by the body in sensation, only by a point of contact between the material and the immaterial which he associated with the pineal gland at the base of the brain.

Now this constitutes presupposition change at a very basic level indeed. Aristotle's pneuma and psyche were placed outside the province of the sciences. Physical reality was to be understood and studied as a great machine. This was to be the province of the sciences, and matters of mind and soul were to be left to the

---

*For this reason, Descartes' followers later criticized Newton for distinguishing *mass* or quantity of matter from the geometrical concept *volume*. As a tradition of physical *theory*, the school of thought epitomized by Descartes was soon overridden by the Galileo-Kepler-Newton tradition, which included ontological presuppositions having to do with physical concepts as well as mathematical ones.

religious authorities. Natural science must be formulated without any mention of ourselves or of God.

In laying out the strict division between mind and soul on the one hand and a mechanical physical reality on the other, Descartes articulated a new set of presuppositions which was already implicit in the work of others and which was to characterize the great synthesis associated with Newton.

To the same extent that the new presuppositions about the purely mechanical character of physical reality formed the background of physical science for the next 300 years, Descartes' conception of the human mind influenced the study of human beings themselves. The human being was to be understood as a body, essentially mechanical in character, and a mind—the self—which was identified with *conscious* thinking. Many earlier natural philosophers had taken it for granted that there were mental factors lying outside of conscious awareness, but influential upon consciousness. Descartes took awareness—thinking—to be the defining characteristic of mind. When the study of human psychology began in earnest some time later, the corrective notion of "unconscious" mind had to be re-invented and distinguished from the conscious thinking substance postulated by Descartes. This is what Arthur Koestler calls the Cartesian catastrophe: the combination of the two doctrines that there is nothing in the mind of which we are not aware and that the mind and the body are forever distinct.[26]

But it seems a bit extreme to call the Cartesian presuppositions a catastrophe, however much we may disagree with Descartes now. It was precisely the separation of mind and body for which he argued that led to the enormous successes of physical science from Newton onward. There was a tightly defined and generally accepted set of presuppositions, a scientific orthodoxy, which took fully 200 years to explore and which broke down only with the coming of relativity theory at the beginning of the twentieth century.

As it became solidified in the great body of theory developed by and because of Isaac Newton, the Cartesian framework carried natural science as far as it could in the attempt to understand nature as something wholly distinct from the human observer—as a mechanism to be analyzed for its structure and internal workings.

Holism and Pluralism

Looking back over the history of natural philosophy up to the time of the great Newtonian synthesis, it becomes clear that the distinction between western and eastern thought which seems so extreme at present is not so firmly rooted in antiquity as one might suppose. Plato and the Pythagoreans, and Descartes as well, tried to view reality as a non-physical abstraction, but the dominant conception of reality in western thought—the one which held sway for the longest period of time—was the organic conception associated with Aristotle. This unitary and essentially holistic view held until the time of Descartes and Newton and the rise of the mechanical view of the universe.

Action at a distance, which had to be explained by talk of mechanical but "incorporeal" agencies like the electric fluid, has always proved problematic to the mechanical world view. In oriental science, at all its various stages, action at a distance was never problematic. If all reality is to be understood as an organic unity, then there is no burning question about *why* the lodestone compasses point to the north. There is a web of relationships throughout the universe, just as there is a web of relationships within the human body. If we can accept the interaction of the human organism without understanding a mechanical connection between willing and acting, between sensing and knowing, then why should we look for mechanical connections in the rest of the organic universe? The effect of the moon on the tides is part of the natural functioning of the organism that is the world.[27]

The perennial philosophy of China has been called "organic materialism" by Joseph Needham. The tendency has never been toward mechanical explanations. Every phenomenon of nature is connected with every other phenomenon of nature; there is order and pattern, which includes the order and pattern of human experience. Objects are not static, independent entities to be studied in a "frozen" state but transitional stages of a living universe. The world is understood as a dynamic network of events, and human experience of what we have come to call "internal" events and "external" events is both the model for, and an inseparable part of, that world.

The idea that all reality, including the "internal" and the "external" aspects of human experience, constitutes an *organic unity* smacks of mysticism to many western thinkers. But on the face of it there is nothing mystical at all about the idea of organic unity. It is, if anything, more economical and straightforward than the separation of the mental from the physical aspects of human experience which dominated western physical science from Newton to the recent past.

The general eastern view, if we follow Northrop's suggestion that there is such a thing, is that all human experience arises out of a relationship between the individual and the one reality of which the individual is himself a part. What modern western "common sense" has tended to distinguish as "mental," and therefore essentially private matters, and "physical," and therefore essentially public matters, are viewed from the East as one and the same thing.

There is a Hindu concept, Maya, sometimes misleadingly translated as "illusion," that Fritjof Capra explains in an intriguing way: Maya is the mistake of confusing a particular point of view on reality—a particular way of sorting out our experience—with reality itself. It is the mistake of "confusing the map with the territory," in Capra's words.[28] It would be as if we were to select one of the three maps of San Francisco from Chapter 3 as the sole correct map or one of the three descriptions of the incident on the beach as the sole correct description.

The dualistic map of the reality of human experience has served western science well. By viewing one whole area of human experience as a great mechanism, separate from human beings themselves, Newton integrated existing theories of planetary motion and the relationships among objects into a tight and well-integrated set of somewhat altered theories in a new context of presuppositions. The key insights of his predecessors in science like Kepler and in natural philosophy like Descartes laid the groundwork for the great integration.

But the mechanical theory of the universe is a map. It is not the territory. Like any map, it can tell us just so much about the territory and no more. Physics has already gone beyond the limits of the Newtonian map of reality, and new maps, with new legends, have appeared in this century. And, what is most important, *we have learned to read the new maps*, whether or not we can read them as well as a professional physicist can.

The characteristic oriental map of reality, similar in many ways to that of Aristotle, is unitary. When we look at the world with this kind of map, we are looking at it from within. The observer himself is included in what he is observing. The very notion of "observation" becomes awkward in this context. The modern western distinction between the observer and the observed does not hold up. This is a different sort of map of reality than the Newtonian or Cartesian map. It is, I suggest, another sort of map that we can learn how to read, as we have learned to read the new maps of contemporary physical science. It is, moreover, a map that is similar in many respects to the new maps of reality which physical science draws.

The Limits of the Mechanical World View

I think that there is a good reason why physics in particular has become a global science, crossing all national and cultural boundaries, in the past fifty years. The penetration of the atomic structure of matter has brought about the downfall of the Newtonian view of the world by pushing it to its limits. The new presuppositions of physics (to whatever extent we can speak at present of there being a consensus on such presuppositions) constitute a shift to a point of view which is essentially holistic rather than pluralistic. The sharp distinction between matter and energy, and even the sharp distinction between observer and observed, have broken down in atomic physics. As many physicists have noted, contemporary physical theory bears striking similarities to the traditional eastern conception of *one* fundamental reality and it is strikingly different from the pluralistic conceptions underlying Newtonian physics.[29]

The door has been opened to "non-materialistic" points of view on the world by the very science which Olmsted described as seeking the laws of the material world. The emphasis I have placed on the long period during which Newton's physics was generally accepted should not lead you to believe that it was immediately enthroned, sanctified, and placed beyond any critical challenge. Newton's system of laws was under constant challenge. It became established as an orthodoxy by meeting the challenge successfully for so long.

At the beginning of the eighteenth century, right after Newton's grand formulation of physical laws, chemistry was already well established as a practical art. Many chemical reactions were well understood, thoroughly predictable, and familiar in ordinary manufacturing, medicinal, and household applications. If the Newtonians believed that they had a *new* understanding which could explain all natural phenomena as the workings of a mechanical system, then it was up to them to show that it could, beginning with the familiar chemical processes.

Here was a theoretical problem, a critical challenge for the overall Newtonian philosophy of nature, and Newtonian physicists addressed themselves to it with energy and enthusiasm. Joseph Black

(1728-1799), Henry Cavendish (1731-1810), Joseph Priestley (1733-1804), Antoine Lavoisier (1743-1794), John Dalton (1766-1844), and many others set about to explain the well-known chemical reactions, to devise experiments which would put their new hypotheses to the critical test, and eventually to arrive at a systematic account of the interactions of the "simple substances" which, according to Newton, must underlie all chemical compounds and reactions. (It was during the critical debate between Priestley and Lavoisier that the "crucial" experiment concerning oxygen described in Chapter 5 was carried out.) Along with metals, sulfur, and bismuth, Lavoisier's first list of "simple substances" included caloric—"the matter of heat"—and light itself.[30]

The critical debate in chemistry went on throughout the eighteenth and nineteenth centuries, and it culminated, so far as the matter of "simple substances" was concerned, with the theses underlying the periodic table of the elements developed by Mendelejeff. Heat, light, and electricity did not, of course, submit to explanation as simple substances or chemical elements. They resisted satisfactory explanation in mechanical terms right through the time of the critical challenges reported by Olmsted in 1851 and even through Maxwell's work with wave functions later in the nineteenth century.

Between Newton and the latter part of the nineteenth century, the activity of the scientific community was very much like the activity in the 1,500 years following Aristotle. Particular areas of theory, with the ontological and categorial presuppositions which delineated their ground, developed within an overriding set of presuppositions, a natural philosophy. The consequences and applications of the theories were spelled out, and new technologies were developed. Mechanical and chemical energy were explained and exploited. Stored energy in the form of chemical compounds could be tapped in ways not previously understood, and the mechanical principles of the Newtonian physics made possible the utilization of that energy in what must forever be known as the "machine age."

But from the time of Maxwell's work in the 1860s and 1870s, the difficulties with the Newtonian view increased. The mechanical view demanded a commitment to *atomism*; there had to be such things as the smallest constituents of matter, and these were presumed to be the atoms of the chemical elements. Energy had to be exchanged among the constituents of matter either in direct mechanical interaction of material objects or as wave energy propagated in a medium. But by the 1890s it became clear from experiments with cathode-ray vacuum tubes and the discovery of X rays and radioactive elements that either there were constituents of matter smaller than the elemental atoms themselves, or else energy was being exchanged in a way other than as continuous waves in a medium.

Less than three weeks before the turn of the twentieth century, Max Planck placed before the critical community a new hypothesis—that radiant energy (the "incorporeals" again) was not exchanged between material objects in smooth, continuous waves. Energy traveled in distinct minimum amounts, or "packets," Planck argued; these were *quanta* of energy, whose minimum size depended upon the frequency of the radiation. In 1905, Einstein took a crucial step further. Not only was energy exchanged between material substances in discrete and discontinuous quanta rather than smooth waves, but energy *existed* in discrete quanta as *photons*, or particles of light. But these, as I noted earlier, were not particles of matter in any sense that could be accommodated in the Newtonian view of nature. Photons could not be understood unequivocally either as "chunks of stuff," which interacted with other chunks in a mechanical way and which had mass, position, and velocity as understood in the Newtonian picture, or as waves propagated through a flexible material substance.

Electromagnetic radiation, which includes light, became the central topic in physics. The term "particle" became a clear misnomer for the quanta of energy being discussed. The limiting presuppositions of the Newtonian natural philosophy had been set aside once and for all. The atom had a *structure*. It was not homogeneous and indivisible. Matter and energy stood on opposite sides of an *equation*. Each attempt to pin down the basic physical constituents of matter has given way to further analyses into constituents which are more basic. At the latest reports, we are down to "quarks," and the end is not yet in sight.* In the developments of quantum mechanics and contemporary physical theory, it becomes clear that "constituents" is as misleading a term as "particle" to use in describing the basis of physical reality. Such terms are

---

*A conception of the atom developed early in the century by Niels Bohr (1885-1962), and long since abandoned as being nowhere near the basic atomic structure from a physical point of view, provides the theoretical base for contemporary chemistry, as described in the section on explanation in Chapter 4.

holdovers in the vocabulary of science from the common sense conceptions of another day.

Newtonian mechanics involved idealized versions of familiar objects: perfectly rigid pendulums, perfectly elastic molecules, perfect and subtle fluids. Space and time really were as we experienced them, and causal relationships on a purely mechanical understanding had to apply to the basic constituents of matter as much as they did to a system of gears and pulleys. This was the great common sense philosophy of nature.

But Newtonian common sense doesn't apply in the subatomic world. There are no fundamental constituents of matter which can themselves be understood as matter. Moreover, as Werner Heisenberg has demonstrated (in his famous "uncertainty" principle), the observer cannot be separated from the observed. The natural philosophy of Descartes and Newton has to be set aside. Any change in observational method entails a modification in the structure of the "particle-process" being observed, and there is no clear-cut way to describe the objects of our observations unequivocally either as particles *or* as processes.

The basis of physical reality must be understood not as a set of static independent objects, but as transitional stages in a dynamic network of events. The "particles" we speak of are "nothing but the shadow of particularization and we cannot ascribe any degree of (absolute or independent) reality to them." The quotation in this last sentence, which might have been written by one of the quantum physicists, was written in the first century A.D. by Ashvagosha, a Buddhist patriarch.[31] On any interpretation of quantum mechanics (and there are several), "natural science must be understood as a part of the interplay between nature and ourselves; it describes nature as exposed to our method of questioning." The quotation in *this* last sentence, which might have been written by Wang-shou-jen in the fifteenth century, was written by Werner Heisenberg in 1958.[32] The convergence I have been talking about between East and West is well underway in the philosophy of nature that underlies contemporary science.

What remains to be said, then, about the Newtonian and Cartesian philosophy of nature and the theories which developed within it? First, it is clear that the common sense presuppositions concerning space, time, and causality stand unquestioned as they relate to human experience. When we set up an observation or an experiment, we presuppose that there is a causal chain of events that leads from a subatomic event to an observer. But the presuppositions which characterize our experience cannot always be assumed to be descriptive of the subatomic event itself.[33] Atomic events are not mechanical events.

Second, when we talk about the physical events which lie within given ranges of size, the mechanical laws formulated by Newton are correct, and they cannot be improved upon.[34]

What Einstein and the quantum physicists have shown is not that Newtonian physics was mistaken, but that there is another physics beyond it. As a physical theory of those systems which can be dealt with as mechanical, the Newtonian conception is correct. But not all systems can be dealt with as mechanical systems. As a *philosophy of nature*, the Newtonian conception is limiting and inadequate. Science has outgrown it.

FURTHER READINGS

Three excellent, detailed, and readable sources in the history of western science are the books by Stephen Toulmin and June Goodfield, *The Architecture of Matter* (1962), *The Fabric of the Heavens* (1961), and *The Discovery of Time* (1965). In addition, *A Short History of Scientific Ideas to 1900* (1959), by Charles Singer, gives a thorough and compact chronicle.

The standard source in the history of oriental science is Joseph Needham, who has a number of books on the subject. See in particular *The Grand Titration* (1969) and *Science and Civilisation in China* (1956). In addition, Carsun Chang's *The Development of Neo-Confucian Thought*, 2 vols. (1962) is excellent.

The connection between contemporary western physics and traditional Indian and Chinese thought is spelled out in a clear and fascinating way in *The Tao of Physics* (1975), by Fritjof Capra. Capra is a physicist and a student of oriental thought.

Those interested in the Greek influences on European thought are encouraged to read Aristotle's corpus, in particular *De Caelo*, and Plato's dialogue *Timaeus*.

I have had to set aside Indian thought here. A good summary of the relevant philosophy is to be found in the introduction to Radhakrishnan and Moore, eds., *A Sourcebook in Indian Philosophy* (1957).

African thought, which was not systematically set down in writing until recently, is covered in several recent works, notably *African Religions and Philosophies* (1969), by John S. Mbiti. F. S. C. Northrop's *The Meeting of East and West* (1946) remains the classic study of the matter, although it is by now somewhat dated.

SEVEN

The Science of Ourselves

Let me come right to the point of this chapter, because the point is already implicit in the last arguments of the preceding one. It seems clear that the recent changes in the presuppositions of physical science have inevitable implications concerning the presuppositions of what I will call *the science of ourselves.* By this, I mean to include, among other fields, those disciplines of the western intellectual tradition which study human beings: medicine, physiology, psychology, sociology, and anthropology. I will confine the discussion to these reasonably well-defined fields for the present, but with a caution: as I argued in Chapter 5, the term "science" cannot usefully be limited to specific subject matters. There is no area of human experience that is not subject to scientific study, and there are very few hypotheses that cannot be addressed scientifically and subjected to public, critical examination.

Admissible Data

One of the difficulties with formulating a science of ourselves in the context of the Cartesian–Newtonian mechanistic view of natural phenomena is that human beings don't seem *to themselves* to be mechanisms of any sort. Attempts to bring the science of ourselves within the mechanistic presuppositions took one of two forms. Either human beings were treated *as if they were* mechanisms of a particularly complicated sort, or else those aspects of human experience which didn't fit within the mechanical picture were set aside as "non-empirical," or "non-scientific," or just not fit for genuine scientific study.

If human beings *are* fit objects for scientific study, then we must be clear concerning just what kind of data must be accounted for by theories about human beings and what kind of evidence can legitimately be brought to bear on hypotheses about human beings.

In Chapter 3 I discussed the matter of establishing a common ground on which critical discourse can take place, and I likened that common ground to the legend of a map. Objective claims can be understood and evaluated only if we understand the legend of the map; the common ground of presuppositions on which descriptions and hypotheses are to be assessed. The first question that must be addressed, if we are to see past the obvious differences between the eastern and the western approaches to knowledge and establish a common ground for the evaluation of hypotheses and theories, is this: What are we to count as admissible data about ourselves? What kinds of statement about human beings *can* be assessed on a common and well-understood ground?

The only defensible answer to this question must refer back to intersubjective agreement—that is, agreement among people, as I argued earlier and as others have argued with different ends in view.[1] Intersubjective agreement depends upon the ability of a number of people to assess a hypothesis or a descriptive claim and arrive at a critical judgment as to its truth or falsity. Such agreement requires communication, and communication requires language—a way of describing events involving human beings that clearly *makes sense* among those using it.

When we disagree on any claim, the standard of objectivity that we must retreat to is based on the inescapable fact that people are pretty much alike, that your experiences and my experiences are similar because you and I are similar. The biological similarities among human beings discussed in Chapter 2 establish the foundation for such claims of similarity and, at the same time, establish the parameters in which the enormously diverse behavior, language, and habits of thought of human beings take place. There are many different maps of human experience—many different ways of describing what we must view as essentially the same *kind* of experience, because all human beings are the same kind of organism. An individual who moves from one part of the world—one culture or cline—to another typically has to learn to read a new kind of map of day-to-day experience. As I noted in the last chapter, we can and do learn to read new kinds of map. We are much more flexible intellectually than the tight doctrines of the mechanistic philosophy of nature would suggest.

Once we are able to agree on a way of describing and discussing our experiences, the similarity among human beings enables us to assess each other's assertions *about* our experiences. We can learn different ways of sorting out those experiences conceptually in order to describe them, in much the same way that we can, on learning a new language, learn to apply new color words and to make new discriminations in the features of our visual experience. There is no need, in laying down criteria for objective judgments, to make a distinction between those experiences which center on objects outside ourselves and those which center *on* ourselves. We can communicate objectively about our emotions, feelings, wishes, desires, and inner states just as well as we can about our visual experience, because we have to make precisely the same assumptions about the overall similarity of human beings in either case.

There is, then, no fundamental difference in discussing "inner" and "outer" experience. What leads us to think that there is a difference is, I suspect, precisely the habits of scientific thought associated with Newton and Descartes that have been ingrained for so long in the language of the western scientific cline. But if you reflect on the fact that it is only the commonality of human experience that lends objectivity to any assertion, the familiar Chinese and Indian doctrine that reality must be experienced, rather than observed from afar, becomes more comprehensible. Moreover, the "mystical" ring of that doctrine is less apparent. The particular experiences we have as individuals are private, but the *kinds* of experience we *can* have are like those of other human beings. The very existence of language depends upon this fact.[2]

There is another way of describing the shift in the presuppositions of science since the breakdown of the mechanical view. From Newton to Einstein, western physical scientists looked at nature in the third person: *that* does such and such, *those things* have such and such properties, or *they* behave in particular ways. As it has become clear in physics that the sharp distinction between the knower and the known does not hold up, there has been a shift in perspective from the third person to the first person—from "they" to "we," in effect. If the differentiation doesn't hold up, then human beings are part of nature; we view it from within, not from without, and we cannot indulge ourselves in the fiction that we *can* set ourselves apart from, and perhaps above, nature. To talk about a particular set of objects is to talk about those objects as they are discriminated and experienced by human beings. To talk about them in a particular way is to construct a map of reality which others can know how to read, either by already understanding the legend of the map we are constructing or by learning to understand it.

What the recent presupposition change in western science makes clear is that science cannot be treated as a spectator sport. We cannot talk about nature without talking about ourselves at the same time, as Heisenberg observed, even when the objects we are talking about are things other than human beings. This is true in two ways. First, in observations about the structure of matter the observer and the observed must be viewed as part of the same system of interrelated events (as discussed at the end of the last chapter); and, second, claims for the objective accuracy of any descriptive assertions about nature must always return to claims about human experience. The demand that experiments must be *replicated*—reproduced by other observers—is an acknowledgement by the scientific community of these very facts.

What happens when the critical dialogue is about ourselves in the more overt sense? There are as many different conceptual points of view possible on human experience of ourselves as there are on human experience of anything else. Agreement about such experience requires communication, and communication requires language. Modern western scientists have typically made the demand that the language of the science of ourselves be entirely limited to third-person assertions. As western scientific action shifted to northern Europe during the seventeenth century, the Cartesian view prevailed: The human body is a machine; the human mind is wholly conscious and involves no events of which the subject · is not consciously aware; and mind and body are forever distinct entities.*

In Chapter 5 I described hypothesis formation, the "bright idea,"

---

*John Locke (1632-1704) and the other British empiricists agreed with Descartes about the conscious nature of mind, holding that it is impossible to perceive without perceiving *that* you perceive. There were dissenters from the beginning: Shaftesbury, John Norris, and others in England; Leibniz, Wolff, and later Immanuel Kant on the Continent. Arthur Koestler reports at length on this controversy in *The Act of Creation*, Chapter 7, and L. L. Whyte at even greater length in *The Unconscious Before Freud*.

the creative act in science, as occurring effectively in random ways against the background of a problem situation. But we know perfectly well that hypotheses don't just occur to us at random. They occur in a context of questions and problems about a subject matter approached from a given point of view, or they may occur to us *about* the point of view itself. Perhaps we are looking at the problem situation in a way that isn't going to lead us anywhere; perhaps if we looked at it another way it would become clear.

We have all had bright ideas about problems at one time or another, and in many cases we are able to tell how they occurred to us: a book we had read the night before on another subject entirely, or a bit of conversation, or perhaps an insight into a personal situation leap to mind as being strongly analogous to a particular perceived problem and suggest an approach to that problem.

But such mental functions as unconscious associations of ideas, or deciding what to think about, or choosing hypotheses to pursue, or, prior to that, wondering about one problem rather than another, cannot be accounted for by any mechanical view of the self, and they have been assumed to be outside the scope of scientific inquiry which is influenced by the Newtonian–Cartesian orthodoxy. Indeed, at times, the very existence of such activities has been denied.

One characteristically orthodox approach to the study of the human self has been the study of human beings as "behaving bodies." Given a stimulus of a certain kind, the human organism responds in a particular and determinate way. How the individual conceives the stimulus, how and why the individual directs his attention toward one object rather than another, are left out of the scientific picture. Such matters as will, volition, and decision are inaccessible to behavioral "third-person" studies and must be set aside.

The aim of such behavioristic studies has been to develop a comprehensive account of human beings in which we are depicted as beings of determinate characteristics, subject to the influence of causal chains of events, like a Newtonian mechanism. The way we respond to stimuli is understood mechanically, in terms of built-in propensities, tendencies, and learning mechanisms. A human being *just is*, on this view, a behaving body. Any claim to an active "inner life"—mental images, trains of thought, ethical decisions, matters of value and priority in arranging the objects of concern, sentimental associations, wishes and conscious decisions—all of the things which make us *human* subjects—are set aside and effectively lost to scientific inquiry.[3]

Behavioristic psychology at its worst studies nothing but the movements of bodies. At its best, it studies, as we did in Chapter 2, a species among other species, a biological system. Biology, too, has been at great pains to conform to the mechanical philosophy of nature. How are we to conceive living beings? Are they distinct from mechanical systems? If so, *how* are they distinct? The third-person approach to science demands that the questions be put in that way.

But we know biological systems in a way that we can never know mechanical systems: firsthand. The human self *is* a biological system, and this can—and need—no longer be understood as a "special kind of complicated mechanism." Once natural selection has developed an organism in which the elements of consciousness are present, the perceptions, needs, desires, and judgments of individuals and clines enter into the evolutionary picture. George G. Simpson, the distinguished biologist on whose writing I drew heavily in Chapter 2 for the account of organic evolution, has this to say about such perceptions, needs, and desires:

> Once they have arisen, *their further evolutionary role is not mechanistically determinate* and is subjected to the influence not only of the actual needs and desires of the group and of volitions extremely complex in basis but also of an even more complex interplay of emotions, value judgments, and moral and ethical decisions.[4]

If there are any causes which operate on human consciousness, they are not all causes of the familiar, mechanical kind, and their manner of operation is unlike the mechanistic necessity of the "gross" physical processes described by Newtonian mechanics. The elementary events involved cannot be broken down further and further to mechanical events of greater and greater complexity. They must be treated as activities of the whole living organism.[5] We respond selectively to stimuli, both consciously and unconsciously.

There are still lessons to be learned from Newtonian physics in the face of all this talk of its breaking down, and of there being only one reality after all. The Newtonian view of nature failed as a view of nature, not as a theory of mechanical systems. To understand reality, we sometimes have to restrict our view of it for a time, back off from it, and break it into pieces in order to develop maps of it. In doing so, we sometimes make the mistake of confusing the map with the territory. As with the three descriptions of the event on the beach in Chapter 3, we have to realize that while many different descriptions of a state of affairs can "tell what really happened," they will tell it from different points of view. No one of them tells the whole story. Indeed, no one account *can* tell the whole story.

The biggest mistakes in science tend to be philosophical mistakes. We draw our maps of reality, and we subject them to critical, objective evaluation, which is the way of science. But often we then go on and take the map to *be* the reality, and this is a mistake.

The reason that western scientists have accepted a limited view for the science of ourselves is that they have tended to consider much of the relevant data as beyond critical evaluation. For example, the familiar first-person experience which underlies rational discourse, our understanding of each other's feelings and behavior, our use and learning of language, the subtle shifts in point of view that we lead each other to in discussing our experience, and even the activity of rational argument itself, are set aside. The behaving-body approach to human experience cannot study the self and its activities, no matter how much insight of a particular sort it may give us into certain public aspects of the self. It cannot even explain how the rational critical dialogue functions, upon which its very own activity as a science depends.

The task for the science of ourselves is not to question the existence of the ethical, emotional, aesthetic, and valuative activities of human beings but to explain them. What special conditions affect the forms that such activities take in one situation or another? How do we learn to understand each other's thoughts as they are expressed in language and as they sometimes defy such expression? How do we communicate our ideas to each other? How do we communicate our emotions? How do we "make each other see" things in given ways so as to establish a common ground for communication?

We have all had the experience of silently wondering what was going on outside our room, of silently trying to see if we *can* imagine a four-dimensional physical object, of silently imagining or wishing during a dull meeting or lecture that we were walking down a sandy beach. Such matters are beyond the scope of any study of behaving bodies. Can we study the silent experiences of human beings? Of course we can. But we can't study them silently. Science depends upon critical dialogue. The questions concerning our silent experiences have to be recognized as genuine scientific questions. What

Heisenberg noted about physics, we must assume about the science of ourselves: that science describes nature—including human nature—as it is exposed by our hypotheses, by the questions we ask, and by the answers we are willing to consider. It is clear that we can't talk about whole human beings without talking about our individual selves as well.

But is it possible to conduct something like a controlled experiment if we adopt this point of view? What constitutes a controlled experiment if the observer and the observed cannot be clearly distinguished?

Not to labor the point, the question of controlled observation and experiment about ourselves must come down to judgments expressed in the first-person plural. We are talking about ourselves, and the relevant data consists precisely of reports about ourselves and our experiences under certain conditions—whether the object of those reports is a reaction that we see taking place in a test tube when we are concerned with the replicability of experiments involving chemical events, or a reaction taking place within our individual selves when we are concerned with the regular consequences of particular sets of circumstances in personal experience. This is as true of experiments involving such matters as visual after-effect as it is of experiments involving particular medicinal substances or procedures, or the effects of self-initiated states of meditation.

Two Approaches to Helping Ourselves

It is impossible to set down in a concise way a natural philosophy which characterizes all of western medicine and psychiatry. The Cartesian division between matter and the conscious mind left the study of medicine with a problem. How is the human self to be viewed, and how are its disorders to be dealt with? The debates in medicine and in psychology since Descartes have centered on these questions: Are there or are there not processes that go on in organisms which have no counterpart in inorganic systems? And, second, Are there or are there not processes that go on in human beings which do not go on in other organisms? The arguments in western medicine have gone back and forth in the centuries since Descartes, and this is not the place to attempt a history of the attitudes.[6] Darwin's work in developing the overall plot of natural selection had an influence on the medicine of the past century, by giving a key to the view that human beings are indeed part of nature, that events in human experience are part of the natural course of events. But the arguments in the science of ourselves haven't been settled, and I am not going to pretend that they can be settled here and now.

Instead, and without attaching any particular names to the positions, let me characterize the kind of approach to the human self which results from a consistent adoption of the Cartesian presuppositions. If medicine and psychology are to be sciences, on this view, they must maintain the overall commitment to the distinction between mind and body and to a mechanistic view of the body. Various suggestions have been raised over the years about the existence of a "vital force" which distinguishes animate matter from inanimate matter. These have been rejected by and large, either because the postulated vital force was too *unlike* the familiar mechanical forces that could be idealized as "subtile fluids" or "minute particles" to be acceptable, or because the postulated vital force was too *similar* to the mechanical forces to succeed in setting animate beings apart from inanimate ones. The retreat from vital forces has almost always been to the position that animate beings, including conscious, willing, animate beings, could be understood fully in terms of complex, interconnected physical—and, therefore, essentially mechanical—processes. Animate beings differ from inanimate ones, in this view, only in their complexity. However, some processes are peculiar to living beings precisely because of their complexity: the regulatory functions, which maintain body temperature, heartbeat, digestion, and so on; and the developmental functions, which govern growth and reproduction, the generation of cells, and the healing of damaged parts of the body.

At its most extreme, the mechanical view of the self has led to what is called "symptom medicine": the specific, mechanically oriented treatment of localized symptoms on a strictly mechanical cause and effect basis. "Symptom medicine" has of course had its psychological counterparts.

Before contrasting this somewhat sketchy view with the general eastern point of view, I should note that the sketch does not characterize either the medicine or the psychology of modern western science in an accurate way, because it has in fact proved impossible for these fields to remain within the Cartesian presuppositions. Despite the frequent relegation of herbal cures to "folk medicine," along with the connection between personality and obvious physical disorder, and matters of diet and personal habit, it has become increasingly plausible to understand habit and personality as being closely interconnected with identifiable bodily states. There are, it seems, certain personality types who are more prone, or less prone, to given diseases than others. Many herbal cures that were viewed as primitive and superstitious by the scientists of a hundred years ago have turned up as effective treatments for specific ills as well as the general maintenance of well-being. Increasingly, medical studies of human populations show that people in specific locations are freer of heart disease, or prostate disease, or cancer, or ulcers than others, or simply that they tend to live longer. Sorting out just which of these characteristics have to do with heredity and which are due to habits of living, diet, characteristic attitudes, trace minerals in the water supply, or other factors is a matter for hypothesis and testing, and it has been increasingly *recognized* as such. The approach to disease and disorder has been increasingly an approach to an understanding of a complete self in its setting rather than an understanding of a mechanical system.

But let me return to that sketch of the mechanical view of the self long enough to contrast it with a characteristically eastern approach to medicine. The typical oriental approach to the self is not at all different from the oriental approach to other matters. The self has been seen as a part of organic nature and has, since the earliest writings available (the *Nei Ching*, set down at least as early as 350 B.C.), been taken as the model on which to understand other matters.

The conception of energy, Ch'hi, is the key to understanding nature in general. In particular, it is the key to the Chinese medical technique called *acupuncture*. Ch'hi is understood to be strongly analogous to the contemporary western conception of a *field* or *energy field*. The presence of matter as individual objects is understood as a rhythmic condensation of the energy which permeates the entire universe.[7] In any particular physical manifestation, the energy/matter strikes a balance between the opposites yin and yang. Particularly in the human self, the equilibrium of Ch'hi is a delicate and shifting one.

This understanding of the objects of experience leads to quite a different set of questions and answers from those of western medicine. Questions about human disorders came to be understood as questions about disturbances of energy in the human system.

According to tradition, the basic data which led to the development of the technique were known as long as 4,000 years ago.[8] The observations have, of course, been refined and modified as the conception of Ch'hi has been refined over the centuries. It seems to have been during the neo-Confucian period of the Ming Dynasty (associated in Chapter 6 with Wang Shou-jen) that the concepts received their present "traditional" form.

Medical observations, like any others, are guided by hypotheses which arise within a context of presuppositions and assumptions. If we presuppose that we are dealing with an organized energy field rather than with a mechanism, we are going to ask different questions about the human self: How does the energy flow? If there is illness, where is the flow disrupted? Where do the disturbances manifest themselves? How can the balance be restored?

The observations consisted of looking for the patterns of flow of energy throughout the body. This was not taken to be the same as the flow of fluids within the body; these are themselves taken to be physical "condensations" of the energy. From the observations, some theses emerged about the human body which stood up to further testing and criticism: When a given bodily organ or function is impaired, certain areas at the surface of the body—areas on the skin—become sensitive to the touch. From one individual to another, the same areas of the skin become sensitive in the event of specific diseases or dysfunction. Of course, this was understood as evidence about how the energy called Ch'hi flowed through the body, and the next reasonable step was to *chart* the flow of the force. What emerged from this tracing of the points of sensitivity on the skin was a map of the human body which showed lines of connection between the points—*meridians* or channels along which the Ch'hi force was understood to flow. Each meridian was named for the organ or function with which it was associated. The meridian associated with the lungs, for example, extends from the upper chest, down through the arm, and terminates in the thumb. Specific *points* are identified along the meridian as the areas of greatest skin sensitivity in the event of lung disorders.[9]

Organs are identified by their function in the oriental understanding of the human self, and they are not in all cases the same identifiable tissue masses within the body which are traditionally distinguished in the West. Organic dysfunction is understood as an imbalance or irregularity in the flow of Ch'hi. Several means, including the use of needles, massage, heat, or small electrical impulses, are used either to stimulate or to retard the flow of energy along the identified meridians in order to affect the relevant functions.

The important question about acupuncture is whether or not it works in predictable ways, and the answer, by all indications, is that it does. Particularly in inducing local anaesthesia, acupuncture has proved effective, and it has received a good deal of attention in the western medical community in recent years in that connection. But acupuncture has also proved effective in treating certain ailments under clinical conditions of observation. It is, of course, not infallible, and it is not a "miracle cure" as the sensational first recent reports seemed to suggest. It is neither magical nor mystical. It is an effective technique for dealing with certain human ills and discomforts that was devised and is understood within a different conception of the nature of the human self than that which has become established in western medicine.

Legitimately, the first reaction of the western medical community has been to attempt to explain the effectiveness of the technique in more familiar terms. The western approach to the human self still tends to look first for a more or less mechanical explanation for bodily effects. If energy of one sort or another follows a specific pattern in the human body, we should be able to see *how* the pattern is followed. Neither the network that transmits the blood nor the network of nerve fibers in the human body follows the course of the acupuncture meridians. One measurable physical correlate with the acupuncture meridians has been reported, however. The electrical resistance of the skin drops appreciably at the points on the acupuncturists' map of the human body.[10]

Here, then, is a problem situation: How and why does acupuncture work? How are the effectiveness of the technique and the measurable changes in skin resistance to be explained? Any hypothesis raised to explain acupuncture will be raised against a background of presuppositions about the nature of the human self and, within that, against a background of assumptions which constitute the body of accepted theory. One set of hypotheses has received serious critical attention concerning the anaesthetic effects of the technique. This is the "gate theory" of pain, which postulates the opening and closing of certain nerve channels under specifiable conditions.[11]

There is, of course, an explanation available from within the oriental presuppositions that define the context within which the technique developed. But it seems unlikely that wholesale presupposition change is going to happen, at least not immediately. What must happen first within a scientific community is that hypotheses will be offered in an attempt to explain the effect within the prevailing presuppositions. The situation is not unlike that in physics just before the turn of the century. There is an article of faith that all phenomena can be explained within a given set of presuppositions, and, until or unless a new set of presuppositions emerges that will explain the successes of medicine as practiced according to both the eastern and the western understandings of the human body, the problem situation will continue.

Two Approaches to Improving Ourselves

I raised the question several pages back whether or not we can formulate hypotheses about our inner states and subject them to critical examination. I think that we can, as I have indicated. Let me illustrate this point and, at the same time, illustrate in another way the difference between the characteristic eastern and western approaches to the self and its states.

Until very recently, the western tradition of medicine and psychology was of the opinion that we had clear-cut answers to questions about just how much conscious, deliberate control we have over our own states. In the past ten years, the answers and the questions have changed.

"Mental functions," wishes, desires, or decisions do not fit easily into the conception of the self as a machine, and claims that seem to challenge the mechanical view have been subjected to considerable testing, with interesting results. Part of the motivation for such testing has been the increasing contact between western and eastern traditions of medicine and psychology. The claim to be investigated is that "inner states"—psychological states, if you will—can straightforwardly produce "outer" results in the overall human self—results which can be approached in "third-person" terms. The means of investigation has been a series of experiments with *biofeedback*. The aim of biofeedback techniques is to bring under conscious control bodily processes and states which are ordinarily thought to be outside such control.

The autonomic nervous system maintains the economy of the

body. It regulates heartbeat, breathing rate, the level of moisture on the surface of the skin (measured by galvanic skin response), peristalsis, brain patterns, and so on. There are typically no clear-cut sensory signals to indicate changes in such bodily functions, unless the changes are well outside normal limits. In fact, in western languages we have no straightforward and scientifically respectable vocabulary for describing such sensations.

The premise of biofeedback techniques is as follows: if a clear and unambiguous signal can be given to indicate when each of the autonomic functions is in a desired state, then the subject may be able to learn to control them. This involves such equipment as sensing devices to monitor the bodily state and communicating devices to produce a signal (typically an audible tone) when the function is within the desired limits. Effectively, one learns to maintain the signal which is linked to the sensing equipment. Thus, for example, if the aim is to maintain the systolic blood pressure below, say, 130, the subject is outfitted with a sensing device to detect the blood pressure. The sensing device is connected to a signal source which produces a tone through a loudspeaker when the blood pressure falls below 130. The "subjective" aim is to maintain the tone.

The difficult thing about biofeedback is that there is no straight-forward technique available for maintaining the signal, and beyond that there is no satisfactory explanation for how and why it works. The instructions to the subject are typically confusing and enig-matic: "Don't try to keep the signal on"—yet the aim is to do just that—"Remember how it feels when the signal is on."[12] But there is no description of where the "feeling" is to be located or how to maintain the "feeling." The cues are subtle enough that they seem to defy description.

Neal Miller, an experimenter in biofeedback, says "It may be that we are not conscious of these sensations *because* we have not been trained to label them."[13] There is no straightforward vocabulary in English for discussing such sensations.

The results of biofeedback experiments have been well publi-cized. What they come down to is that the use of biofeedback equipment to bring bodily states under conscious control has not proved very reliable, although it has demonstrated that conscious control of bodily states is possible. After startling initial successes in a given group of experiments, the results typically take a sudden turn and become highly variable. And, oddly enough, this has been the case even when laboratory animals, rather than human beings, have been the experimental subjects (rewarded with food or pleasurable sensations when able to maintain the signals).[14]

With human beings, biofeedback techniques have proved difficult or impossible to take out of the laboratory. A subject who shows marked success in maintaining a steady alpha-wave pattern, or in keeping blood pressure or heartbeat patterns in some desired configuration in the laboratory, is not as a rule able to repeat the laboratory successes in a normal day-to-day routine without the gadgetry.

What differentiates the eastern and western approaches to such matters is partly in the point of view—what we take ourselves to be doing when we try to manipulate our own states—and partly in the consequences of the point of view. In many cases, western researchers have adhered to the "conservative" view of the self as a behaving body. Where western researchers hesitate to put a name to the introspective data concerning biofeedback, for fear of prejudicing the experiment, eastern traditions have not only named the states but have, through centuries of discussing them, developed reasonably straightforward techniques for manipulating them. The vocabulary for describing the states is elaborate and full of conundrums. It is, indeed, a different mode of thought, in which riddles often replace straightforward exposition. But the aim is not exposition. The aim is to induce particular states in others and in oneself. The *technology* does not require machines; it requires techniques and people. It is as straightforward as it can possibly be, and it works.

Meditative techniques, in India and Japan in particular, are approached as a matter of personal hygiene. There are a number of different techniques, each having its partisan followers, and there are hypotheses about the relative merits of the various techniques which bear critical examination. Although these techniques have come into the forum of the critical scientific community of the West with what must be regarded as new hypotheses surrounding them, they have grown out of the tradition of the Vedic writings of India, which are roughly contemporary with the Old Testament, beginning as oral traditions and gradually being set down in writing between 1500 B.C. and 500 A.D. The "inner technology" and its accompanying theory are not new, then, except in the sense that they are new to the global scientific community as plausible hypotheses.

The particular inner technology I want to describe here in contrast to biofeedback techniques is one which has come to be known as transcendental meditation. This particular technique has been taught widely over the past few years, and it has been subjected more to critical and experimental tests than have its competitors.

What differentiates transcendental meditation and the other techniques from biofeedback is that they are taught *as* techniques, in much the same way that other skills—like playing a musical instrument—are taught. The student is given a particular sound—a "nonsense word"—which is chosen for him by the instructor after a brief interview. This sound is the *mantra* or *mantram*. The claim is made that distinct personality types respond well to different sounds in this situation, and critical examination of the claim seems to indicate that it is correct.[15]

The technique seems absurdly simple. Basically, one simply sits in a comfortable position, closes one's eyes, and "thinks the sound" of the mantra. That, essentially, is all there is to it, although there are bits of advice, reassurance, encouragement, and refinement of the technique which are passed along from instructor to student in the training sessions—again, rather like learning to play a musical instrument. The technique, and others like it, are a sort of "kindergarten yoga"—the simplest first step in a series of much more complicated techniques which have their origins in the same Vedic traditions. But kindergarten or not, it *is* a technique to be applied by the subject himself, where biofeedback has none, and it has stood up well to the challenge of criticism and experimentation. People who have learned the technique have been tested physiologically before learning it, after six months of practicing it regularly (twice daily for twenty minutes at a time), and during meditation. The results are impressive in support of the thesis stated by two researchers that the meditative "mental states can markedly affect physiologic function."[16]

Moreover, the effects of meditation are different from those of sleep, simple relaxation, or hypnosis. All available physiological measures of relaxation and rest indicate that meditation is effective in lowering blood pressure, tension (as measured by galvanic skin response and blood lactate), and metabolic rate, and in maintaining electrical patterns in the brain associated with a relaxed state. The long-range effects, measured by physiological and psychological tests, seem to indicate that the general efficiency in reaching goals and the physical and psychological well-being of the subject are enhanced.[17] Moreover, during meditation, a peculiar pattern of brain waves (as displayed in an electroencephalogram) is detected. A strong "alpha" wave is associated with a relaxed state in individuals. During meditation, not only is the alpha-wave pattern strong, it is also typically synchronous or rhythmically related to other brain-wave patterns. This pattern seems to be peculiar to the meditative state and is not exhibited during sleep, work, ordinary waking states, or dreaming sleep.

The traditional aim of meditation is somewhat different from that of the biofeedback researches. Meditation, as a technique for enhancing one's own well-being, is viewed as a matter of personal hygiene, as I noted. The stated aims include relaxing and relieving oneself of stress, increasing one's happiness, efficiency, creativity, and insight into oneself and the world at large, and "tuning up" the total self. These are matters which cannot even be sensibly discussed in the vocabulary of behaving bodies, but they are, as I suggested earlier, precisely the areas which a science of ourselves must ultimately attempt to deal with.

An interesting thing about the transcendental meditation technique in particular, among the many related techniques which have been taught recently in western communities, is that it has been deliberately introduced in the West simply *as* a technique, without any well-identified theory stated in order to explain it. It thus comes into the arena of western critical discussion not as a theoretical development from another tradition but as a technique with measurable effects which developed within such a theoretical tradition and which presents a challenge to western conceptions of the self.

Convergence

I suggested earlier that the task of a science of ourselves must be to explain the ethical, emotional, aesthetic, and valuative activities of human beings. Beyond this task, there is a need for self-understanding, in what we distinguish as medical matters and psychological matters as well as in the other areas which I listed at the beginning as being parts of the overall science of ourselves. We need an understanding both of our individual selves and of humankind in general. There are needs, tasks, and problems which bear down on human beings in all times and in all places which are not—cannot be—outside the province of the critical community which I have identified as scientific.

I do not think that what I have identified as the eastern view of the self could or should be adopted in the West in its entirety, including the particular ontological and categorial presuppositions which characterize the background against which acupuncture and meditation were developed as practical techniques. We can certainly learn to "read the maps" of the traditional Chinese philosophy of nature concerning yin and yang and the Ch'hi energy, and it has become fashionable to do so at particular times and places. We could, perhaps, understand the more directly theistic theoretical background against which meditative techniques were developed. But I do not think that this is appropriate. What *is* appropriate is a critical challenge to the traditional presuppositions of both the eastern and the western conceptions of the self, and it seems to me that there is at present a crisis in the science of ourselves which is very much like the crisis in physics at the beginning of this century.

The *holistic* view in physics—essentially that matter and energy cannot ultimately be distinguished—defies understanding in terms of Newtonian common sense. But beyond having reports of laboratory experiments which bear out the holistic view on critical grounds, we have had dramatic demonstrations of practical developments which arose from within the point of view and which bear it out in very public ways. If any single event marked the breakdown of the mechanical view of the universe for the community at large beyond the scientific community, it was the release of the films of the first atomic bomb explosion in 1945.

In the science of ourselves, there has not been such a single, dramatic event to mark and emphasize the shift away from mechanistic conceptions of the self. What has happened instead is that another conception of the self has been recognized as analogous in some ways to the new conceptions in physics. The inevitable popular sensation created by such a new development has pretty much come and gone. Practical consequences of the holistic view of the self—essentially that mind and body cannot ultimately be distinguished—lack the public dramatic impact of an atomic bomb explosion. The consequences are emerging gradually as it becomes apparent that this view of the self has resulted in the development of effective techniques for helping and improving ourselves. Despite some attempts to develop a holistic conception of the self during the development of psychoanalytic theory, the theoretical shifts are only beginning to happen, and the convergence of the two views of the self is only just getting underway. But it seems indisputable that as the scientific community has developed hypotheses for approaching nature, nature has forced us to the holistic view in physics, and that nature is forcing us to the same kind of view in the science of ourselves.

FURTHER READINGS

In *Knowing and Acting* (1976), particularly in the final chapter, Stephen Toulmin spells out in some detail what he considers to be the coming task for psychology and philosophy in developing a new approach to the science of ourselves.

A particularly good constructive critique of western psychology is Floyd Matson's *The Broken Image* (1966).

A good place to start in the study of acupuncture is Manaka and Urquhart, *The Layman's Guide to Acupuncture* (1972).

On transcendental meditation, see Bloomfield, Cain, Jaffe, and Kory, *T.M.: Discovering Inner Energy and Overcoming Stress* (1975).

A general summary of recent movements in nonstandard self-help and medical techniques, with critical observations on each, is Adam Smith's lively and entertaining *Powers of Mind* (1975). More detailed study is to be found in Charles T. Tate's *Altered States of Consciousness* (1969).

EIGHT
One Science

In Chapter 5, I suggested that there is virtually no hypothesis that is itself unscientific. What is properly judged to be scientific or unscientific is the way in which we address hypotheses. I think that this is a fair estimation of the current open attitude of the overall scientific community in the wake of the relatively recent turn away from a mechanical philosophy of nature. The former limiting presuppositions of natural science, which considered natural phenomena to be only those that could (potentially or actually) be explained as the workings of mechanical systems, and which drove the sharp Cartesian distinction between the mental and the physical deep into the common sense idioms of the community at large, have been largely abandoned.

It may well be that the presuppositions of a scientific community at a given time are so imbedded in the idioms of that community that they remain relatively hidden while they are current, emerging only in retrospect or when they are under direct challenge. Nevertheless, I think it is fair to say that at present the overall scientific community is relatively free from limiting presuppositions, at least those presuppositions which limit what is and what is not to be understood as a natural phenomenon. Ontological presuppositions—about the sorts of entity that can exist—have changed rapidly and repeatedly in physics over the past fifty years, rapidly and repeatedly enough that they may never again settle down to a firm commitment that there is *nothing but* certain sorts of entity. There is, however, a clear methodological demand that hypotheses be subjected to the disciplined critical dialogue of the scientific community against the background of well-established theoretical beliefs and commonly experienced phenomena.

Any theory within the overall context of scientific beliefs has, of course, its own *warranting* presuppositions which give sense to assertions about entities of given kinds and the properties that they have. But, especially in the area of atomic physics and in the new approaches to the science of ourselves, there is often a difficulty in sorting out entities from properties of entities or from properties of entire systems, in the sense that the distinction between entities and their properties is usually understood.

If we understand the question whether or not a given subject matter is scientific to be the question whether or not a comprehensive critical method can be applied to it, we extend the scope of what is to be understood as the proper purview of science, and we extend it in two directions. On the one hand, it is reasonable to hold that logic, geometry, and mathematics have been brought into the area of science and do not constitute an independent area of study that holds a special status apart from natural science itself. Given the alternative geometries which began to develop during the nineteenth century, the complex and sometimes controversial developments in formal logic of this century, and the alternative approaches to the understanding and manipulating of arithmetic and higher mathematics, it has become clear that the acceptance of a mathematical or logical proof is as much a matter of critical evaluation as is the acceptance of a hypothesis in natural science. There are questions to be answered critically about what is to count as a proof, as well as questions about methodology.[1] There have even been cases where a proof of a given logical or mathematical result was generally accepted by the relevant community and later shown to be fallacious.[2]

On the other hand, some aspects of human experience that were regarded as unscientific according to the mechanical world view have come to be seen as fair game for scientific investigation. The increasing popular and scientific interest in what were regarded in the past as hidden experiences of human beings—the *occult*, literally—can be understood as an expression of dissatisfaction with the recent history of western science and the view of ourselves that resulted from it. If the behavioral approach to psychology is understood to lead to the claim that the human self *just is* a behaving body, there is good reason to be dissatisfied. In our own experience of ourselves, we know that the claim is false.

The newly broadened view of nature and of the scope of science has led to a general questioning of the notion that some events are to be considered "supernatural" and outside the scope of systematic inquiry. But the very notion of "supernatural" events seems nonsensical. Some natural events are within human control and some aren't. Some can readily be explained and some can't. But to say that something is beyond control or beyond present comprehension is not to say that it is outside the natural course of events.

I think it fair to say that the current scientific attitude holds that there are no supernatural phenomena at all, only some that don't submit readily to certain kinds of explanation. Such an attitude doesn't preclude a healthy skepticism about whether or not certain kinds of "occult" phenomena actually do take place, nor does it preclude seeking to explain unusual events in familiar terms. But contemporary scientists are less likely to dismiss out of hand the possibility of events which cannot be easily explained, and there is no longer the facile dismissal of "non-mechanical" events as illusory. If something happens, and we can establish that it does, it is natural. The question then becomes, how are we to understand it? This attitude in itself raises the level of scientific curiosity, and it broadens the scope of just what is to count as a scientific question.

In particular, the recent broadening of scientific curiosity has led to new questions about human beings and their experience. We want a more satisfying conception of the self than the mechanistic view provides. What has happened is that earlier conceptions of what it is to be a human being and of the things that human beings can do have been revived for a second critical look. Extrasensory perception, hypnotic healing, healing by interpersonal forces or influences, rhythms of human intellectual, emotional, and physical activity ("biorhythms") have all become respectable fields of inquiry and are addressed systematically within the scientific community. I don't mean to suggest that the scientific community has come to believe flagrant claims in all of these areas but, rather, that hypotheses about all of these areas are currently under critical discussion and testing within the mainstream of the scientific community. A reasonable barometer of the attitude of a scientific community at large is the activity of its professional associations. The American Association for the Advancement of Science voted to include the Parapsychology Association among its affiliates in 1969.

It is well known that both Sigmund Freud and Carl Jung believed in the occurrence of telepathic communication between patient and analyst and that Jung went on to attempt to give a systematic explanation of extrasensory perception and other parapsychological phenomena.[3] But, in large measure, these interests of psychological theorists, as well as the beliefs of Newton, Kepler, and other physical scientists in astrology or psychic phenomena, were regarded until quite recently as nonscientific quirks of great scientists.

What Is Scientific?

When we reject one account of ourselves which doesn't seem to give an accurate picture of human experience, we want to avoid just latching on to the next conception of the human self which happens along. The recent great interest in views of the self that are and have been held by other cultures and communities can lead to a good deal of shopping around for a view of ourselves that seems congenial. But this is best done in a critical way, and the historical scholars are contributing to contemporary psychology in ways that they have not done before. Ancient views of the self and the influences on it have been trotted out for a second critical look after having been regarded as mere superstition during the period of the mechanical world view.[4]

There is an interesting connection between the new respect for the beliefs of the past and the critical process as I have described it here. I think we have come to realize that beliefs which were widely held at a given time must have been held for a reason. They didn't just arise spontaneously. If the critical activity directed toward arriving at group solutions to common problems really does date back to the time of the earliest human communities, then there must have been a critical context in which any given body of generally held beliefs made sense and stood up to competing hypotheses. Beliefs are shared, expressed, and criticized in language, within an idiom, on a common ground of presuppositions. If we understand the common ground, we can better understand and evaluate the beliefs.

Some of the ancient beliefs are easily explained away once we understand the context in which they developed and why they came to be held. We can understand the diverse beliefs in local nature gods without being tempted to adopt any of them. We can also understand the concepts of heaviness and lightness in Aristotle's

physics from the perspective of contemporary physics, and see
perfectly well why, on his understanding of motion, when you stop
pushing something, it stops moving.

But other beliefs which were set aside by the scientific community
under the influence of the mechanical world view are not so easy to
explain away, and these bear further examination. As I suggested at
the opening of Chapter 1, the lid does seem to have come off the
trashbin into which mechanistic science tossed many beliefs of the
past. The intellectual climate today is right for a further critical
investigation of those beliefs.

Perhaps there is something to be learned from what were regarded as unscientific views about psychic or parapsychological
phenomena—extrasensory perception, telepathy, clairvoyance, and
psychokinesis. Claims for individual "special talents" in these areas
have been made throughout history, and so have claims about other
influences on human personality and activity which do not operate
in any obvious manner.[5] Astrology and divination have played
important roles in political and military decisions of the past. People
really believed that there was guidance to be had in these ways, and
many people still believe it. Some of the claims of "special powers"
are clearly fraudulent or misguided, but others are not clearly so.
The difficulty comes when we try to sort out one from the other, and
it is worth doing, as many scientists have come to agree. The
distinguished physicist Gerald Feinberg has posed two questions
which he believes will be clarified by research into psychic phenomena: "What is the range of consciousness in the universe? How is
consciousness related to other aspects of the world?"[6]

But despite the general willingness to consider such matters, the
lid isn't off. We don't want to believe just anything. Rational beliefs
must stand up to systematic criticism and must fit together to form a
coherent picture of reality. We want to distinguish between new
insights or insights which may be expressed in unfamiliar idioms,
and the ramblings of the charlatan who makes things up as he goes
along.

Individual people, and even groups of people, may hold beliefs
for reasons having to do with their own psychological characteristics, or simple, honest ignorance, more than with the content of the
beliefs themselves. The "cargo cults" of New Guinea come to mind
in this connection. During World War II, an airstrip was cleared in
New Guinea so that supplies could be delivered to troops in the
area. Local natives saw the planes come and go, and they noticed

that when the lights on the airstrip were turned on at night, cargo planes appeared out of the sky bearing gifts for the people on the ground. Long after the war ended, and the troops left, and the airstrip was abandoned, the natives maintained vigils at the airstrip and lit fires at night to bring the gift-bearing vehicles back. They clearly believed that it was the activities on the ground which attracted the gifts. One hopes that by now someone has had the decency to land on that airstrip one last time and tell the people to go home.

Surely, we can sort out those beliefs which are rooted in such honest ignorance from those which are worth critical consideration, and we can sort out psychological explanations about why people hold given beliefs from questions about whether or not there might be something to them.

The Criticism of Minimal Hypotheses

Let me recall the schema of selection one last time, without presenting it diagrammatically. Against the general background of established scientific beliefs, I want to suggest how hypotheses of legitimate scientific interest can be framed about extrasensory perception, and "occult" phenomena in general.

A "new" hypothesis that is considered by a global scientific community may not actually be very new at all, but may have originated in the ancient beliefs of one or another of the cultures which comprise the overall community. When such hypotheses are to be examined outside the local idiom in which they arose, they have to be extracted from that context and laid out conservatively as *minimal* hypotheses against the general background of established scientific beliefs. In this way, beliefs about the efficacy of meditation were stripped down to the bare bones claim that "mental states can markedly affect physiologic function" by the western researchers who began the current inquiries.

We can formulate correspondingly minimal hypotheses concerning both psychic phenomena and astrology, and critically evaluate them in straightforward ways. If such hypotheses stand up to critical examination and experiment, further explanatory hypotheses will not be far behind. There will be a clear need to incorporate the hypotheses into the overall scientific picture by explaining them.

A minimal hypothesis concerning extrasensory perception is the following:

> There can be significant correlation between two people's
> mental states or thoughts without contact through the
> known senses.

Research is underway now at facilities throughout the world, trying
to evaluate hypotheses which are more elaborate than this one and
which involve commitment to specific sorts of phenomena. But a
minimal hypothesis of this sort is the beginning point for such an
inquiry. It is fully plausible, according to the criteria developed
earlier. Beyond that, after more than thirty years of research and
testing, many researchers regard it as established to a statistically
significant degree.[7]

What is missing is an explanation. There are not at present any
further well-established hypotheses which would serve to tie in
extrasensory perception with other phenomena by explaining how
such events can take place. Several explanations have been offered,
postulating particles which are exchanged between functioning
human beings,[8] or postulating a property ("factor") of both space
and consciousness which can influence and be influenced by human
thought.[9] Neither the particle nor the factor seems to have clear-cut
bearing on other well-understood phenomena, however, and no
explanation is at present generally accepted.

There are some minimal hypotheses to be extracted from the
claims of astrology, which seem to be amenable to straightforward
examination in the light of evidence. Here is one, absolutely minimal, that should be explored as a prelude to any further critical
examination of the area:

> The time of year in which an individual is born is correlated
> with characteristics of the individual's personality.

This one should be easy enough to explore, given the many sorts of
personality test and inventory that have been devised and adminis-
tered to large segments of the population over the years in schools
and in military service. I am not suggesting, of course, that any
existing test provides a full picture of human personality, but there
are enough measurements in such tests of how an individual reacts
to given situations and problems so that, if any correlation exists, it
should appear. If all the personality tests that are filed away were
sorted out by birth date, and it turned out that indeed there were
discernible patterns of common traits in people born at given times
of the year, it would support the hypothesis as I have stated it.

But even the strongest confirmation of this hypothesis would not lead us to astrology. If such similarities among people with common birth dates do exist, they could be explained by seasonal or other factors as well. This is only the first step.

Here is a second hypothesis relevant to astrology, expressed in minimal terms:

> The position of the earth relative to other bodies in the solar system is correlated with features of human personality and emotions (and therefore human events).

Before any serious inquiry into the claims of traditional astrology can be carried out, a minimal hypothesis like this one must be examined for its initial plausibility and subjected to critical examination in the light of available evidence. As with the hypothesis about extrasensory perception, there seems to be no problem with plausibility. An argument for the plausibility of this hypothesis would follow essentially the same lines as the assessment of the levitation hypothesis carried out in Chapter 3, with the exception that there *do* seem to be clear-cut instances or precedents of the effects on human emotions of relationships between the earth and other bodies in the solar system. There is fairly well-established evidence that the phases of the moon have an effect on human emotions, particularly in situations where people are confined in hospitals, penal institutions, or even boarding schools; and psychiatric studies indicate correlations between the emotional states of patients and the time of the full moon. In addition, recent studies have indicated that there are distinct and regular rhythms in people's emotional states that may be correlated with the relative positions of the earth and other bodies in the solar system.[10]

The initial plausibility of the two minimal hypotheses about astrology does not, however, confer plausibility on the detailed and complicated claims associated with traditional astrology as it has come down from Egyptian, Chaldean, and medieval times. It does not even confer plausibility to the more circumspect assertion that the stars influence human behavior. But suppose that the second hypothesis is understood, not merely as something having to do with the full moon and people's moods, but as a more general assertion that the position of the earth relative to *all* the bodies of the solar system has an effect on the individual's personality and thus on the events of his life. Surely this hypothesis is subject to critical evaluation through the study of many thousands of cases, analyzing the planetary positions at the time of a person's birth as compared to details of his or her biography.*

Suppose that there does turn out to be a correlation between the positions of the planets at the time of an individual's birth and features of his personality and biography. Suppose that the discernible rhythms of physical, emotional, and intellectual activity in a person's life do turn out to be correlated with the state of the solar system at the moment of his birth. What will have been established? Surely not the elaborate doctrines of the ancient soothsayers, but, at most, support for a natural hypothesis that the gravitational patterns that exist in the solar system at any given time are related to the activities of living beings. We know this, in minimal form, already. Sea animals whose changing activities were thought to be related directly to the local tides turn out to be affected by the position of the moon instead. How was this determined? By moving them inland (to Illinois) and observing their activities in laboratory situations. The rhythms of their activities adjusted after a short time to what the tidal situation *would have been* if Illinois were under water, as calculated from the position of the moon and its known effects on the tides in a given area.[12] What hasn't been established is the extent to which gravitational patterns affect the activities of human beings, or how subtle those influences might be, or whether or not it might be gravity, after all, which figures in extrasensory perception rather than a new particle or "factor."

I am not suggesting that any of these plausible hypotheses or interesting correlations would establish the traditional tenets of astrology. What I am suggesting is that any body of beliefs that has been held for a long time bears serious and careful critical examination. Beliefs persist for a reason, and while the reason isn't always that they are true, it also isn't always that they are simply comforting beliefs for people to hold.

A critical approach to bringing together the traditional beliefs of different human clines entails looking at both the idioms and the specific claims of distinct human traditions of knowledge and belief. If there is anything to a given set of claims, we should be able to ferret it out in the critical process. If there isn't anything to them, we

---

*Such an analysis was begun during the 1930s by Karl Ernst Krafft (1900–1945), whose work was interrupted during World War II and never resumed. The results were indefinite.[11]

are bound to find out interesting and useful things along the way as we establish that there isn't.

Science and Religion

Neither of the opposing views in the perennial and tiresome conflict between simplistic approaches to science and simplistic approaches to theology seems to reflect any sense at all of the distinction between a map and the territory it represents. In either physical science or theology, any claim that we have the final insight at a particular time must be dismissed as insufferable arrogance. Insights can come from within any idiom. Insights into the meaning and purpose of human existence are as hard to come by as insights into the nature of physical reality, and we must take our insights where we can find them.

There is a long history to the conflict between religion and science, but I can see little present justification for it. The ancient nature gods served to fill the gaps in physical theory precisely because of the personalities that were postulated for them. *How* they ruled nature was of little concern; it could be summed up as the magic of the gods, essentially outside human capability and understanding. But as new philosophies of nature developed, relationships were discovered among events in human experience that could be understood without any direct reference to superhuman personalities. If the gods did intervene in human affairs, their intervention had to be understood as less direct than had been supposed. The gaps in physical theory were filled by more physical theory. Each solution to a problem raised new problems and suggested new solutions. The magic of the gods, if it existed at all, was not *simple* magic, and the personalities became less and less well defined in religious traditions.

The gradual development of human science to its present global scale has been accompanied by bitter and sometimes bloody controversy over the centuries. Part of this can be dismissed as the result of entrenched religious institutions or of the combinations of religious and political power in the same institutions. Perhaps we can identify particular heroes or villains in the history of science and established religion, but it is hard to identify particular *arguments* on either side of the conflict which hold any credibility.

On the one hand, as more and more detailed explanations of natural events have developed, many people have come to believe that the conceptions of religion are no longer meaningful. All the nature gods, they argue, have been explained away, and with the nature gods went any notion of meaning or purpose to human existence.

On the other hand, many people take the opposite view: all their sense of right and wrong and of the meaning of human existence is bound up with a set of beliefs that seems to be directly contradicted, and therefore menaced, by the development of scientific explanations. If our values are to stand, then we must reject science, they say.

Neither of these lines of argument is valid. The choice, it seems to me, is not between a mechanical world without meaning or purpose and a world run by the magical stunts of a wizard. Surely, there is nothing in a belief in a god that is going to conflict with attempts to come to an understanding of nature, nor is there anything in scientific theory that is going to disprove the existence of a god or lead us to the conclusion that human existence is meaningless and ethical concepts vacuous.

The understanding of the world and of ourselves that is the object of scientific activity cannot be ignored when we seek a further understanding of the meaning and purpose of human activity and of appropriate goals for such activity. At the same time, the meaning, purpose, and goals that we associate with our activity in general cannot be ignored when it comes to doing science. It would seem that the more insight we have in one area, the more we are likely to have in the other.

Now it may appear that I'm hedging here. On the one hand I'm talking about religion and on the other hand about meaning, purpose, and goals, with no apparent connection to any particular set of religious beliefs. But that is just the point of my argument. Insights can be developed and expressed in many different idioms, as I have argued throughout this book. If they come from a particular religious tradition, so be it. What is untenable is any claim that a good insight constitutes the final word. The greatest blasphemy against an infinite deity is not to question whether or not it exists, but to claim that all its characteristics are known and understood at any particular time. Particular conceptions of reality should not be confused with reality itself. Maps should not be confused with the territory they represent, whether they be theological maps or scientific ones.

The End of Science

I have laid out a theory about scientific activity in this book and have spelled out some of the arguments for, and consequences of, the theory. In closing, I want to reflect briefly on some of the implications of the view I have presented.

As the many different environments of human beings are being drawn together into one global environment by the rapid and detailed communication of ideas and beliefs, it seems clear that the many individual beliefs and traditions of belief that arose in these separated environments are being drawn together into one global community of science. The recent international and intercultural character of physics seems to be just the beginning of the development of a science that can only be identified as human science rather than as eastern or western. But I do not think that this suggests that there is to be some final formulation of theory about the nature of the world that will be drawn from all available insights and constitute the last word of ultimately correct scientific theory. There are several reasons to doubt that science will end in such a way.

First, it has become clear that there is not a uniquely correct way of describing our experiences. There are ways that are appropriate and testable in given contexts of discussion, and ways of relating our observations to specific theories or problems. We never start from neutral observations, as Karl Popper points out, but always from problems. We approach problems by asking questions and offering hypotheses. Questions are asked from a point of view, and hypotheses are conjectured from a point of view. And as I have argued, there is no such thing as a uniquely correct point of view.

Beyond the matter of understanding the point of view from which questions are asked, hypotheses framed, and experience described, there is the matter of proof. Hypotheses are tentative solutions to problems. They are subjected to criticism and ultimately to experiment, both of which are intended not to demonstrate them as true in some final sense but to *prove* them in the sense discussed earlier in Chapter 4—to test them by attempting to refute them. When a hypothesis is accepted as a thesis of a nascent or existing theory, it is because it has stood up to rational and experimental criticism better than its competitors have. The process is strongly analogous to the natural selection of organisms, and I have suggested, quite seriously, that there is more than an analogy involved.

Our knowledge consists, at any given time, of those beliefs which have shown their comparative fitness by withstanding the criticism of the community of scientists to which they are proposed. New hypotheses are never proposed in a vacuum. They are offered against the background of an existing set of beliefs. The ecological niche into which they are to fit is never empty, and a new hypothesis must prove itself more fit than the established beliefs.

A scientific theory can never be fully justified, and it cannot be demonstrated to be true in some final sense, apart from the context of presuppositions in which it is proposed. It can be criticized and tested; and if it stands up, the most that can be said for it is that, after the criticizing and testing, it seems more interesting, more powerful, more promising, and better able to explain what is taken to be the relevant data than its competitors. Its truth is always to be judged within its own context.

Even this degree of success for a theoretical hypothesis—when it has reached the point of being the best available theory to explain the data within its scope—guarantees it nothing in the long run but a place in the history books. The very problems which a given theory raises, coupled with the constantly changing external problem situation, guarantees that the frame of reference in which problems are described and solutions proposed will change over a period of time. Presuppositions about what sorts of entity can exist, and about what sorts of property those entities can have, will always be subject to replacement.

What is fairly certain about such entities as quarks, gluons, and gravitons is that eventually they will be dropped from our theories, not because what is being said about them is false or nonsensical but because we will eventually "stop talking that way." That is what happened to phlogiston and ether and the electric fluid. There is nothing in current physical theory that entails their non-existence. The context in which we describe and explain the relevant events has no place for discussing them at all. Statements about phlogiston and ether can no longer be assessed for truth or falsity in the context of physics. We simply don't talk that way any more.

The important sense in which we have to say that the current theories of physical science are not ultimately true is the historical sense. The sorts of entity and property we are willing to talk about in our explanatory theories are constantly changing. Nature's laws are not expressed in some unique way that can be identified as nature's terms, because nature's terms don't exist as something independent from a community of knowing subjects who seek to understand nature.

We can learn to read new maps of reality. We have recently done so in physics, and we are presently doing so in the science of ourselves. The overall similarity of human experience makes it possible for us to learn to read each other's maps. Whether or not human beings can learn to understand reality in non-human terms remains an open question. It is fascinating to suppose that we might discover that dolphins do indeed have languages as complex as human languages and that we might go on to discover how to understand such languages. Overcoming the differences in sensory experience and the concomitant differences in conceptual organization between men and dolphins will be more difficult than has been supposed. As technology makes it possible for human beings to travel away from this planet, the further question arises whether or not we will encounter other intelligent species that differ from us in more drastic ways than the dolphins do, and whether or not those differences can be overcome and yet another set of maps of reality understood. The fact that such questions arise indicates that science is not about to come to an end because all its problems are solved. There is no end to the problems.

Science might one day come to an end, of course, if we wipe ourselves out. We are all familiar with the frightening scenarios whereby the human species might become extinct through accident, misjudgment, or irrationality on the part of a physicist, a chemist, a biologist, or a politician. It may come to an end because we, as a species, systematically poison ourselves and the environment we live in.

But if none of those scenarios comes to be, the activity of doing science will go on indefinitely. There is no end in sight, no matter how strongly we may feel at any given time that there are "just a few more details to work out." What this means for the natural scientist today is essentially what it meant for Franklin, Du Fay, Newton, Descartes, Aristotle, and Pythagoras. It is the same for theoreticians in any field, including the field which gave rise to the theory in this book: The best we can hope for, in the long run of history, is that we will prove to have been wrong in interesting ways.

FURTHER READINGS

Colin Wilson develops an informed and detailed critique of western science in his recent book *The Occult* (1973), in which he argues for a further consideration of traditional and primitive beliefs. Arthur Koestler, in *The Roots of Coincidence* (1973), develops a set of hypotheses to explain extrasensory phenomena.

A reasonably good summary view of current research in extrasensory perception, psychic healing, and related matters is to be found in Mitchell, ed., *Psychic Exploration* (1974), a collection of articles by researchers on the subject. Blair's *Rhythms of Vision* (1976) is a somewhat more partisan, but quite informative, source for work in biorhythms and for review of particular areas where there are well-confirmed phenomena that cannot at present be explained by physical and psychological theory.

NOTES

CHAPTER 2

1. Charles Darwin, *The Origin of Species*, introduction by Julian Huxley (New York: Mentor, 1958).
2. Karl R. Popper, *Objective Knowledge: An Evolutionary Approach* (Oxford: Oxford Univ. Press, 1972), pp. 242 ff.
3. See, for example, Robert Ardrey, *African Genesis: A Personal Investigation into the Animal Origins and Nature of Man* (New York: Atheneum, 1961), pp. 68–69. Also, see Colin Wilson, *The Occult* (New York: Vintage, 1973), p. 126.
4. David L. Hull, *Philosophy of Biological Science* (Englewood Cliffs, N.J.: Prentice-Hall, 1973), Chap. 2, esp. p. 68.
5. Stephen J. Gould, "Evolution's Erratic Pace," *Natural History* (May 1977), 14–16; Bryan Clarke, "The Causes of Biological Diversity," *Scientific American* (August 1975), 55–60; A. J. Cain, *Animal Species and Their Evolution* (1954; rpt. New York: Harper Torchbooks, 1960), pp. 134 ff.
6. Cain, p. 143.
7. Gould, "Evolution's Erratic Pace," pp. 14–16.
8. John Buettner-Janusch, *Origins of Man: Physical Anthropology* (New York: Wiley, 1966), pp. 498–516, esp. p. 506.
9. George Gaylord Simpson, *Principles of Animal Taxonomy* (New York: Columbia Univ. Press, 1962), p. 91.
10. Clarke, "The Causes of Biological Diversity," pp. 55–60.
11. See Buettner-Janusch, pp. 329–332. See also Adam Smith, *Powers of Mind* (New York: Ballantine, 1975), pp. 330–333; and G. Hugh Begbie, *Seeing and the Eye: An Introduction to Vision* (Garden City, N.Y.: Doubleday Anchor, 1973), pp. 165 ff. and pp. 194 ff.
12. For a brief summary of these questions as they relate to the development of human beings, see Jacob Bronowski, *The Ascent of Man* (Boston: Little, Brown, 1973), esp. Chap. 1.
13. More details on relative visual acuity appear in Begbie, Chap. 5; Conrad G. Mueller and Mae Rudolph, *Light and Vision* (New York: Time-Life Books, 1966), pp. 16–29; and G. Adrian Horridge, "The Compound Eye of Insects," *Scientific American* (July 1977), 108–120.
14. See Joseph Margolis, *Persons and Minds* (Dordrecht: Reidel, 1977); see also P. F. Strawson, *Individuals: An Essay in Descriptive Metaphysics* (Garden City, N.Y.: Doubleday Anchor, 1963), Chap. 3.

15. See Begbie, p. 52 and Chapter 5.
16. B. F. Skinner, *About Behaviorism* (New York: Vintage, 1974), p. 48.
17. Mueller and Rudolph, pp. 126–127; Begbie, pp. 109–141.
18. Dan I. Slobin, *Psycholinguistics* (Glenview, Ill.: Scott, Foresman, 1971), p. 124.
19. See Popper, *Objective Knowledge*, pp. 238–239; Buettner-Janusch, p. 346.
20. Theodosius Dobzhansky, *Mankind Evolving: The Evolution of the Human Species* (New Haven: Yale Univ. Press, 1962), pp. 212–213.
21. Skinner, *About Behaviorism*, p. 50.
22. Clarke, pp. 55–60.
23. See E. O. Wilson, *Sociobiology* (Cambridge, Mass.: Harvard Univ. Press, 1975); Stephen Jay Gould, *Ever Since Darwin* (New York: Norton, 1977), pp. 251–267; and Richard Currier, "Sociobiology: The New Heresy," *Human Behavior* (November 1976), 16–22.
24. Clarke, pp. 55–60.
25. Skinner, *About Behaviorism*, pp. 44, 51.
26. Ibid., p. 64.
27. Dobzhansky, pp. 212–213.
28. Konrad Z. Lorenz, *King Solomon's Ring* (New York: Crowell, 1952), Chap. 8.
29. Dobzhansky, p. 22; Buettner-Janusch, pp. 337–338.
30. Skinner, *About Behaviorism*, p. 141.
31. Julian Huxley, "Clines: An Auxilliary Taxonomic Principle," *Nature*, 142 (1938), 219; Simpson, p. 179; and Ernst Mayr, "Notes on Nomenclature and Classification," *Systematic Zoology*, 3 (1954), 86–89.
32. Cain, pp. 149–150.
33. Slobin, Chap. 5, esp. pp. 123–131.
34. See Slobin, Chap. 5; also, refer to Noam Chomsky, *Language and Mind* (New York: Harcourt Brace Jovanovich, 1968), esp. the early chapters.
35. Skinner, *About Behaviorism*, pp. 48–50.
36. P. F. Strawson, *Individuals: An Essay in Descriptive Metaphysics* (Garden City, N.Y.: Doubleday Anchor, 1963), esp. pp. 168–182.

CHAPTER 3

1. See P. F. Strawson, *Individuals: An Essay in Descriptive Metaphysics* (Garden City, N.Y.: Doubleday Anchor, 1963), pp. 204–210.
2. B. F. Skinner, *About Behaviorism* (New York: Vintage Books, 1974), pp. 48–49. See also G. A. Svechnikov, *Causality and the Relation of States in Physics* (Moscow: Progress Publishers, 1971), pp. 52 ff.
3. Clifford M. Will, "Gravitation Theory," *Scientific American* (November 1974), 25–33.
4. George Gaylord Simpson, *Principles of Animal Taxonomy* (New York: Columbia Univ. Press, 1962), pp. 18–19.
5. See Karl R. Popper, *The Logic of Scientific Discovery* (New York: Harper Torchbooks, 1959), p. 423.

CHAPTER 4

1. The point was suggested by Charles Reed.
2. See Marshall Spector, "Theory and Observation," *British Journal for the Philosophy of Science*, 17 (1966), 1-20 (Part 1) and 89-104 (Part 2).
3. George Gaylord Simpson, *Principles of Animal Taxonomy* (New York: Columbia Univ. Press, 1962), pp. 16-18.
4. J. R. Partington, *A Short History of Chemistry* (New York: Harper Torchbooks, 1965), pp. 342-352.
5. A. J. Cain, *Animal Species and Their Evolution* (1954; rpt. New York: Harper Torchbooks, 1960), p. 19.
6. The DNA molecule of a mammal contains billions of "bits" of genetic information encoded as the sequence of nucleotides in the molecule. A simple DNA molecule, containing 5,375 nucleotide "bits" has recently been fully mapped. See John C. Fiddes, "The Nucleotide Sequence of a Viral DNA," *Scientific American* (December 1977), 55-67. See also Clifford Grobstein, "The Recombinant-DNA Debate," *Scientific American* (July 1977), 22-33.
7. See W. V. Quine, "Two Dogmas of Empiricism," in *From A Logical Point of View* (Cambridge, Mass.: Harvard Univ. Press, 1953), pp. 20-46.
8. See Richard Zaffron, "Identity, Subsumption, and Scientific Explanation," *The Journal of Philosophy* (1971), 849-860.
9. See Marshall Spector, "Theories and Models," *British Journal for the Philosophy of Science* (1965), pp. 121-142; also, see Mary B. Hesse, *Models and Analogies in Science* (Notre Dame, Ind.: Univ. of Notre Dame Press, 1966), pp. 7-57.

CHAPTER 5

1. See "From Reckoning to Writing" (unsigned), *Scientific American* (August 1977), 58 for an account of some recent discoveries.
2. Denison Olmsted, *A Compendium of Natural Philosophy* (New Haven: S. Babcock, 1851), Preface.
3. Olmsted, p. 238.
4. Olmsted, pp. 264-265.
5. Stephen Toulmin and June Goodfield, *The Architecture of Matter* (New York: Harper & Row, 1962), p. 258.
6. Olmsted, p. 270.
7. Olmsted, p. 271.
8. Olmsted, p. 271.
9. David Blair, *An Easy Grammar of Natural and Experimental Philosophy* (Philadelphia: Kimber & Sharpless, 1840), pp. 136 ff.
10. Olmsted, pp. 239, 272.
11. Ibid., p. 272.
12. Ibid., p. 272
13. Ibid., p. 273.
14. Ibid., p. 273.

15. Ibid., p. 274.
16. Ibid., pp. 274–275.
17. Sir Edmund Whittaker, *A History of the Theories of Aether and Electricity* (1910; rpt. New York: Harper Torchbooks, 1960), Vol. 1, p. 363.
18. Toulmin and Goodfield, *The Architecture of Matter*, pp. 249–260.
19. J. Dorman Steele, *Fourteen Weeks in Natural Philosophy* (New York: Barnes, 1876), p. 291.
20. See Brewster Ghiselin, *The Creative Process* (New York: Mentor Books, 1955); also, Arthur Koestler, *The Act of Creation* (New York: Macmillan, 1969).
21. A.C. Crombie, *Medieval and Early Modern Science* (Garden City, N.Y.: Doubleday Anchor, 1959), Vol. 2, pp. 75–84; also, Stephen Toulmin and June Goodfield, *The Fabric of the Heavens* (New York: Harper & Row, 1961), pp. 165–169.
22. William Manchester, *The Glory and the Dream*, Vol. 1 (Boston: Little, Brown, 1973).
23. Toulmin and Goodfield, *The Architecture of Matter*, pp. 194–200 and p. 256.
24. See J. Bronowski, *The Ascent of Man* (Boston: Little, Brown, 1973), pp. 244–256; see also Toulmin and Goodfield, *The Fabric of the Heavens*, pp. 253–256.
25. Clifford M. Will, "Gravitation Theory," *Scientific American* (November 1974), 25.
26. Will, p. 26.
27. Will, p. 28.
28. Will, p. 33.
29. Here my debt to Karl Popper becomes explicit. See his *Objective Knowledge: An Evolutionary Approach* (Oxford: Oxford Univ. Press, 1972), pp. 238 ff. and pp. 248 ff.

CHAPTER 6

1. Werner Heisenberg, *Physics and Philosophy* (New York: Harper & Row, 1958), p. 187.
2. F. S. C. Northrop, *The Meeting of East and West* (New York: Collier, 1966), pp. 312–315.
3. Immanuel Kant, *The Critique of Pure Reason*, A660 and B688 (rpt. New York: Modern Library, 1958), p. 305.
4. Robert Madden, James D. Muhly, and Tamara S. Wheeler, "How the Iron Age Began," *Scientific American* (October 1977), 122–131.
5. For a description of this period in the West, see Stephen Toulmin and June Goodfield, *The Architecture of Matter* (New York: Harper & Row, 1962), pp. 25–40; for the East, see L. Carrington Goodrich, *A Short History of the Chinese People*, 4th ed. (New York: Harper Torchbooks, 1969), pp. 14–17.
6. See Wilfred Sellars, "Philosophy and the Scientific Image of Man," in

Robert Colodny, ed., *Frontiers of Science and Philosophy* (London: Allen & Unwin, 1964), pp. 35–78.

7. Stephen Toulmin and June Goodfield, *The Fabric of the Heavens* (New York: Harper & Row, 1961), p. 56.

8. Arthur Koestler, *The Act of Creation* (New York: Macmillan, 1969), pp. 255–261.

9. See Fritjof Capra, *The Tao of Physics: An Exploration of the Parallels Between Modern Physics and Eastern Mysticism* (Berkeley: Shambhala, 1975), p. 116.

10. For a more detailed account of Plato's theory of matter, see Toulmin and Goodfield, *The Architecture of Matter*, pp. 75–82.

11. Aristotle, *De Caelo*, IV.3 (311B).

12. See Toulmin and Goodfield, *The Fabric of the Heavens*, pp. 94–96.

13. Charles A. Singer, *A Short History of Scientific Ideas to 1900* (Oxford: Oxford Univ. Press, 1959), pp. 62–102; also, Toulmin and Goodfield, *The Fabric of the Heavens*, pp. 90–114 and *The Architecture of Matter*, pp. 82–91 and pp. 109–118; and Koestler, pp. 226–228.

14. Singer, pp. 103–136.

15. Capra, p. 103; also, Goodrich, p. 67.

16. Joseph Needham, *The Grand Titration: Science and Society in East and West* (London: Allen & Unwin, 1969), p. 317. (Needham's translation of this passage, slightly paraphrased and abridged.)

17. Needham, p. 311.

18. Goodrich, pp. 31–35.

19. See, for example, Goodrich, pp. 10–14; Capra, pp. 103–104; and Northrop, pp. 315–317.

20. See, for example, Singer, pp. 137 ff.

21. Singer, pp. 137–187.

22. Koestler, p. 228.

23. Needham, pp. 39–45.

24. Carsun Chang, *The Development of Neo-Confucian Thought* (New York: Bookman Associates, 1962), Vol. 2, pp. 44–60.

25. See René Descartes, *The Philosophical Works of Descartes*, Elizabeth S. Haldane and G. R. T. Ross, trans. (New York: Dover, 1955), Vol. 1, p. 107.

26. L. L. Whyte, *The Unconscious Before Freud* (Garden City, N.Y.: Doubleday Anchor, 1960), p. 25; also, Koestler, pp. 148–149.

27. Needham, pp. 21 ff.

28. Capra, p. 88.

29. Heisenberg, pp. 76–92; also, J. Robert Oppenheimer, *Physics and the Common Understanding* (New York: Simon & Schuster, 1964), pp. 8 ff.; and Capra, pp. 130–160.

30. Toulmin and Goodfield, *The Architecture of Matter*, p. 219.

31. Capra, p. 292.

32. Heisenberg, p. 81.

33. Ibid., pp. 90–92.

34. Ibid., p. 97.

CHAPTER 7

1. See Henryk Skolimowski, "Karl Popper and the Objectivity of Scientific Knowledge," in Paul Arthur Schilpp, ed., *The Philosophy of Karl Popper* (La Salle, Ill.: Open Court, 1974), Vol. 1, pp. 483–508. Also, see Karl Popper, *The Logic of Scientific Discovery* (New York: Harper Torchbooks, 1959), pp. 44–47.
2. See Arthur Koestler, *The Act of Creation* (New York: Macmillan, 1969), pp. 271–300.
3. Ibid., pp. 154–157; also, see Stephen Toulmin, *Knowing and Acting: An Invitation to Philosophy* (New York: Macmillan, 1976), pp. 265–284.
4. George Gaylord Simpson, *The Meaning of Evolution* (New York: Mentor, 1951), pp. 141–142.
5. See Floyd W. Matson, *The Broken Image* (Garden City, N.Y.: Doubleday Anchor, 1966), pp. 143–145; See also Ludwig von Bertalanffy, *Problems of Life* (New York: Harper Torchbooks, 1960), p. 197.
6. But see Stephen Toulmin and June Goodfield, *The Architecture of Matter* (New York: Harper & Row, 1962), Chap. 14; See also David L. Hull, *Philosophy of Biological Science* (Englewood Cliffs, N.J.: Prentice-Hall, 1974), Chap. 5.
7. Fritjof Capra, *The Tao of Physics: An Exploration of the Parallels Between Modern Physics and Eastern Mysticism* (Berkeley: Shambhala, 1975), pp. 213–215.
8. Yoshio Manaka and Ian A. Urquhart, *The Layman's Guide to Acupuncture* (New York: Weatherhill, 1972), p. 22.
9. Manaka and Urquhart, p. 43.
10. Charles Panati, *Supersenses* (New York: Quadrangle, 1974), p. 92.
11. See Ronald Melzack, *The Puzzle of Pain* (New York: Basic Books, 1973).
12. Adam Smith, *Powers of Mind* (New York: Ballantine, 1975), pp. 92–100. My own experiences with biofeedback bear out Smith's descriptions.
13. Smith, p. 102.
14. P. A. Obrist et al., eds., *Cardiovascular Psychophysiology* (Chicago: Aldine, 1974).
15. B. C. Glueck and C. F. Stroebel, "Biofeedback and Meditation in the Treatment of Psychiatric Illness," *Comprehensive Psychiatry* (July–August 1975), 314–316.
16. Robert K. Wallace and Herbert Benson, "The Physiology of Meditation," *Scientific American* (February 1972), 85–90.
17. H. H. Bloomfield, M. P. Cain, D. T. Jaffe, and R. B. Kory, *T.M.: Discovering Inner Energy and Overcoming Stress* (New York: Dell, 1975), pp. 90–107.

CHAPTER 8

1. A particularly problematic kind of mathematical proof has become current with the increase in the use of computers. See, for example, Kenneth Appell and Wolfgang Haken, "The Solution of the Four-Color-Map Problem," *Scientific American* (October 1977), 108–112.

2. For a good general discussion, see Imre Lakatos, *Proofs and Refutations* (Cambridge: Cambridge Univ. Press, 1976). A good example of a logical proof gone awry because of hidden presuppositions is A. N. Prior's "Modality and Quantification in S5," *Journal of Symbolic Logic*, (1956), 60–62. Prior's mistake is caught by Saul Kripke in "Semantical Considerations in Modal Logic," *Acta Philosophica Fennica* (1963), 83–94.

3. See Arthur Koestler, *The Roots of Coincidence* (New York: Vintage, 1973), pp. 82–110 for a concise explanation of Jung's views and of Jung's dialogues with Freud and Pauli on the subject. See also C. G. Jung, *Synchronicity* (Princeton: Princeton Univ. Press, 1973).

4. See Colin Wilson, *The Occult: A History* (New York: Vintage Books, 1973), Part 3; also, see Lawrence Blair, *Rhythms of Vision* (New York: Schocken, 1976).

5. Wilson, pp. 176–199.

6. Gerald Feinberg, in Edgar D. Mitchell and John White, eds., *Psychic Exploration* (New York: Putnam's, 1974), Foreword.

7. Koestler, *The Roots of Coincidence*, Chap. 1; Mitchell, Part 1.

8. Koestler, ibid., pp. 50–81.

9. Wilson, pp. 574–579.

10. Blair, Chap. 1; Wilson, pp. 247 ff.

11. Wilson, p. 251.

12. Blair, p. 54; Frank Brown, "Living Clocks," *Science*, December 4, 1959, pp. 1535–1544.

BIBLIOGRAPHY

Achinstein, Peter, and Stephen Barker, eds. *The Legacy of Logical Positivism*. Baltimore: Johns Hopkins University Press, 1969.

Appell, Kenneth, and Wolfgang Haken. "The Solution of the Four-Color-Map Problem." *Scientific American* (October 1977), pp. 108–121.

Ardrey, Robert. *African Genesis: A Personal Investigation into the Animal Origins and Nature of Man*. New York: Atheneum, 1961.

Baker, Adolf. *Modern Physics and Antiphysics*. Reading, Mass.: Addison-Wesley, 1970.

Begbie, G. Hugh. *Seeing and the Eye: An Introduction to Vision*. Garden City, N.Y.: Doubleday Anchor, 1973.

Benson, Herbert, John F. Beary, and Mark P. Canol. "The Relaxation Response." *Psychiatry*, 37 (February 1974).

Berrill, N. J. *Man's Emerging Mind*. New York: Dodd, Mead, 1955.

Blair, David. *An Easy Grammar of Natural and Experimental Philosophy*. Philadelphia: Kimber & Sharpless, 1840.

Blair, Lawrence. *Rhythms of Vision: The Changing Patterns of Belief*. New York: Schocken, 1976.

Bloomfield, Harold H., M. P. Cain, D. T. Jaffe, and R. B. Kory. *T.M.: Discovering Inner Energy and Overcoming Stress*. New York: Dell, 1975.

Bloomfield, Harold H., and Robert B. Kory. *Happiness: The T.M. Program, Psychiatry, and Enlightenment*. New York: Simon & Schuster, 1976.

Bronowski, Jacob. *The Ascent of Man*. Boston: Little, Brown, 1973.

Brown, Frank. "Living Clocks." *Science* (December 4, 1959), pp. 1535–1544.

Buettner-Janusch, John. *Origins of Man: Physical Anthropology*. New York: Wiley, 1966.

Burtt, Edwin Arthur. *The Metaphysical Foundations of Modern Science*. 1932; rpt. Garden City, N.Y.: Doubleday Anchor, 1954.

Cain, A. J. *Animal Species and Their Evolution*. 1954; rpt. New York: Harper Torchbooks, 1960.

Capra, Fritjof. *The Tao of Physics: An Exploration of the Parallels Between Modern Physics and Eastern Mysticism*. Berkeley: Shambhala, 1975.

Chang, Carsun. *The Development of Neo-Confucian Thought*. New York: Bookman Associates, 1962.

Chomsky, Noam. *Language and Mind*. New York: Harcourt, Brace and World, 1968.

——. *Problems of Knowledge and Freedom*. New York: Fontana, 1972.

Clarke, Bryan. "The Cause of Biological Diversity." *Scientific American* (August 1975), pp. 55-60.

Colodny, Robert, ed. *Frontiers of Science and Philosophy.* London: Allen & Unwin, 1964.

Crombie, A. C. *Medieval and Early Modern Science,* 2 vols. Garden City, N.Y.: Doubleday Anchor, 1959.

Currier, Richard. "Sociobiology: The New Heresy." *Human Behavior* (November 1976), pp. 16-22.

Darwin, Charles. *The Origin of Species.* Intro. by Julian Huxley. 1859; rpt. New York: Mentor, 1958.

Descartes, René. *The Philosophical Works of Descartes.* Trans. Elizabeth S. Haldane and G. R. T. Ross. 2 vols. New York: Dover, 1955.

Dobzhansky, Theodosius. *Mankind Evolving: The Evolution of the Human Species.* New Haven: Yale Univ. Press, 1962.

Eckhart, Ludwig. *Four-Dimensional Space.* Trans. Arthur L. Bigelow and Steve M. Slaby. Bloomington: Indiana Univ. Press, 1968.

Eysenck, H. J. *Sense and Nonsense in Psychology.* 1957; rpt. New York: Penguin, 1972.

Feyerabend, Paul. *Against Method: Outline of an Anarchistic Theory of Knowledge.* Atlantic Highlands, N.J.: Humanities Press, 1975.

Fiddes, John C. "The Nucleotide Sequence of a Viral DNA." *Scientific American,* December 1977, pp. 55-67.

Fodor, J. A., and J. J. Katz, eds. *The Structures of Language: Readings in the Philosophy of Language.* Englewood Cliffs, N.J.: Prentice-Hall, 1964.

Gamow, George. *Thirty Years That Shook Physics: The Story of Quantum Theory.* Garden City, N.Y.: Doubleday Anchor, 1964.

Ghiselin, Brewster. *The Creative Process.* New York: Mentor, 1955.

Ghiselin, Michael T. *The Triumph of the Darwinian Method.* Berkeley: Univ. of California Press, 1969.

Glueck, Bernard C., and Charles F. Stroebel. "Biofeedback and Meditation in the Treatment of Psychiatric Illness." *Comprehensive Psychiatry* (July-August 1975), pp. 314-316.

Goodrich, L. Carrington. *A Short History of the Chinese People,* 4th ed. New York: Harper Torchbooks, 1969.

Gould, Stephen Jay. *Ever Since Darwin: Reflections in Natural History.* New York: Norton, 1977.

———. "Evolution's Erratic Pace." *Natural History* (May 1977), pp. 14-16.

Greene, Judith. *Psycholinguistics: Chomsky and Psychology.* Baltimore: Penguin Education, 1972.

Grobstein, Clifford. "The Recombinant-DNA Debate." *Scientific American* (July 1977), pp. 22-33.

Gustafson, Donald F., ed. *Essays in Philosophical Psychology.* Garden City, N.Y.: Doubleday Anchor, 1964.

Heisenberg, Werner. *Physics and Philosophy: The Revolution in Modern Science.* New York: Harper Torchbooks, 1958.

———. *Physics and Beyond.* New York: Harper & Row, 1971.

Hempel, Carl G. *Aspects of Scientific Explanation and Other Essays in the Philosophy of Science.* London: Collier-Macmillan, 1965.

Hesse, Mary B. *Models and Analogies in Science.* Notre Dame, Ind.: Univ. of Notre Dame Press, 1966.

Horridge, G. Adrian. "The Compound Eye of Insects." *Scientific American* (July 1977), pp. 108-120.

Hull, David L. *Philosophy of Biological Science.* Englewood Cliffs, N.J.: Prentice-Hall, 1974.

Huxley, Julian. "Clines: An Auxiliary Taxonomic Principle." *Nature* (1938), p. 219.

Jarvie, I. C. "Explaining Cargo Cults." In *Rationality,* ed. Bryan Wilson. Oxford: Basil Blackwell, 1970.

Jung, Carl G. *Synchronicity: An Acausal Connecting Principle.* Trans. R. F. Hull. Princeton, N.J.: Princeton Univ. Press, 1973.

Kant, Immanuel. *Critique of Pure Reason.* Trans., with an Introduction, by Norman Kemp Smith. 1786; rpt. New York: The Modern Library, 1958.

Koestler, Arthur. *The Act of Creation.* Danube Edition. New York: Macmillan, 1969.

———. *The Roots of Coincidence: An Excursion into Parapsychology.* New York: Random House-Vintage, 1973.

Kraus, Bertram S. *The Basis of Human Evolution.* New York: Harper Torchbooks, 1964.

Kripke, Saul. "Semantical Considerations in Modal Logic." *Acta Philosophica Fennica* (1963), pp. 83-94.

Kuhn, Thomas S. *The Structure of Scientific Revolutions.* Chicago: Univ. of Chicago Press, 1962.

Lakatos, Imre. *Proofs and Refutations: The Logic of Mathematical Discovery.* Cambridge: Cambridge Univ. Press, 1976.

Lakatos, Imre, and Alan Musgrave, eds. *Criticism and the Growth of Knowledge.* Cambridge: Cambridge Univ. Press, 1970.

Lana, Robert E. *Assumptions of Social Psychology.* New York: Appleton-Century-Crofts, 1969.

Lilly, John Cunningham. *The Mind of the Dolphin: A Non-Human Intelligence.* New York: Avon, 1969.

Lorenz, Konrad Z. *King Solomon's Ring: New Light on Animal Ways.* New York: Crowell, 1952.

Madden, Robert, James D. Muhly, and Tamara S. Wheeler. "How the Iron Age Began." *Scientific American* (October 1977), pp. 122-131.

Manaka, Yoshio, and Ian A. Urquhart. *The Layman's Guide to Acupuncture.* New York: Weatherhill, 1972.

Manchester, William. *The Glory and the Dream: A Narrative History of America, 1932-1972.* 2 vols. Boston: Little, Brown, 1973.

Margolis, Joseph. *Persons and Minds.* Dordrecht: D. Reidel Publishing Co., 1977.

Matson, Floyd W. *The Broken Image: Man, Science and Society.* Garden City, N.Y.: Doubleday Anchor, 1966.

Mayr, Ernst. *Animal Species and Their Evolution.* Cambridge, Mass.: Harvard Univ. Press, 1963.

———. "Notes on Nomenclature and Classification." *Systematic Zoology* (1954), pp. 86-89.

——. "The Nature of the Darwinian Revolution." *Science* (1972), pp. 981–988.

Mbiti, John S. *African Religions and Philosophies.* Garden City, N.Y.: Doubleday Anchor, 1969.

McNeill, D. *The Acquisition of Language: The Study of Developmental Linguistics.* New York: Harper & Row, 1970.

Mead, Margaret, ed. *Cultural Patterns and Technical Change.* New York: Mentor, 1955.

Melzack, Ronald. *The Puzzle of Pain: Revolution in Theory and Treatment.* New York: Basic Books, 1973.

Mitchell, Edgar D. *Psychic Exploration: A Challenge for Science.* Ed. John White. New York: Putnam's, 1974.

Mueller, Conrad G., and Mae Rudolph. *Light and Vision.* New York: Time-Life Books, 1966.

Needham, Joseph. *Science and Civilisation in China.* Cambridge: Cambridge Univ. Press, 1956.

——. *The Grand Titration: Science and Society in East and West.* London: Allen and Unwin, 1969.

Northrop, F. S. C. *The Meeting of East and West.* 1946; rpt. New York: Collier, 1966.

Obrist, P. A., et al., eds. *Cardiovascular Psychophysiology.* Chicago: Aldine, 1974.

Olmsted, Denison. *A Compendium of Natural Philosophy.* New Haven: S. Babcock, 1851.

Oppenheimer, J. Robert. *Science and the Common Understanding.* New York: Simon & Schuster, 1964.

Panati, Charles. *Supersenses: Our Potential for Parasensory Experience.* New York: Quadrangle, 1974.

Papazian, Haig P. *Modern Genetics.* New York: Signet, 1967.

Partington, J. R. *A Short History of Chemistry.* New York: Harper Torchbooks, 1965.

Pauling, Linus. *College Chemistry.* San Francisco: W. H. Freeman, 1957.

Popper, Karl R. *The Logic of Scientific Discovery.* New York: Harper Torchbooks, 1959.

——. *Conjectures and Refutations: The Growth of Scientific Knowledge.* New York: Basic Books, 1962.

——. *Objective Knowledge: An Evolutionary Approach.* Oxford: Oxford Univ. Press, 1972.

Price, H. H. *Thinking and Experience.* Cambridge, Mass.: Harvard Univ. Press, 1953.

Prior, A. N. "Modality and Quantification in S5." *Journal of Symbolic Logic* (1956), pp. 60–62.

Progoff, Ira. *The Death and Rebirth of Psychology.* New York: Julian Press, 1956.

Quine, W. V. *From a Logical Point of View.* Cambridge, Mass.: Harvard Univ. Press, 1953.

Radhakrishnan, Sarvepalli, and Charles A. Moore, eds. *A Sourcebook in Indian Philosophy.* Princeton, N.J.: Princeton Univ. Press, 1957.

Rosenberg, Jay. *Linguistic Representation.* Dordrecht: D. Reidel Publishing Co., 1974.

Ross, James Bruce, and Mary Martin McLaughlin, eds. *The Portable Renaissance Reader.* New York: Viking Press, 1953.

Sagan, Carl. *The Dragons of Eden: Speculations on the Evolution of Human Intelligence.* New York: Random House, 1977.

Sarton, George A. *A History of Science.* Cambridge, Mass.: Harvard Univ. Press, 1959.

Schilpp, Paul Arthur, ed. *The Philosophy of Karl Popper.* 2 vols. La Salle, Ill.: Open Court, 1974.

Searle, J. R. *Speech Acts: An Essay in the Philosophy of Language.* Cambridge: Cambridge Univ. Press, 1959.

Simpson, George Gaylord. *The Meaning of Evolution.* New York: Mentor, 1951.

———. *Principles of Animal Taxonomy.* New York: Columbia Univ. Press, 1962.

Singer, Charles A. *A Short History of Scientific Ideas to 1900.* Oxford: Oxford Univ. Press, 1959.

Skinner, B. F. *Science and Human Behavior.* New York: Macmillan, 1953.

———. *About Behaviorism.* New York: Random House–Vintage, 1974.

Slobin, Dan I. *Psycholinguistics.* Glenview, Ill.: Scott, Foresman, 1971.

Smith, Adam. *Powers of Mind.* New York: Ballantine, 1975.

Smith, Alan G. R. *Science and Society in the Sixteenth and Seventeenth Centuries.* New York: Harcourt Brace Jovanovich, 1972.

Snyder, D. Paul. *Modal Logic and its Applications.* New York: Van Nostrand, 1971.

Spector, Marshall. "Theory and Observation." *British Journal for the Philosophy of Science,* parts 1 and 2, 17 (1966), pp. 1–20 and 89–104.

———. "Models and Theories." *British Journal for the Philosophy of Science* (1965), pp. 121–142.

Stebbing, L. Susan. *Philosophy and the Physicists.* New York: Dover, 1958.

Steele, J. Dorman. *Fourteen Weeks in Natural Philosophy.* New York: A. S. Barnes, 1876.

Strawson, P. F. *Introduction to Logical Theory.* London: Methuen, 1952.

———. *The Bounds of Sense.* London: Methuen, 1966.

———. *Individuals: An Essay in Descriptive Metaphysics.* Garden City, N.Y.: Doubleday Anchor, 1963.

Svechnikov, G. A. *Causality and the Relation of States in Physics.* Moscow: Progress Publishers, 1971.

Swartz, Robert J., ed. *Perceiving, Sensing, and Knowing.* Garden City, N.Y.: Doubleday Anchor, 1965.

Tart, Charles T. *Altered States of Consciousness.* New York: Wiley, 1969.

Toulmin, Stephen. *Foresight and Understanding: An Inquiry into the Aims of Science.* New York: Harper Torchbooks, 1963.

———. *Knowing and Acting: An Invitation to Philosophy.* New York: Macmillan, 1976.

Toulmin, Stephen, and June Goodfield. *The Fabric of the Heavens.* New York: Harper & Row, 1961.

——. *The Architecture of Matter*. New York: Harper & Row, 1962.

——. *The Discovery of Time*. New York: Harper & Row, 1965.

Veith, Ilza, trans. *The Yellow Emperor's Classic of Internal Medicine*. 1949; rpt. Berkeley: Univ. of California Press, 1966.

Verrill, A. Hyatt, and Ruth Verrill. *America's Ancient Civilizations*. New York: Putnam's-Capricorn, 1967.

von Bertalanffy, Ludwig. *Problems of Life*. New York: Harper Torchbooks, 1960.

Wallace, Robert K. "Physiological Effects of Transcendental Meditation." *Science* (1970), pp. 1751–1754.

Wallace, Robert K., and Herbert Benson. "The Physiology of Meditation." *Scientific American* (February 1972), pp. 85–90.

Watson, James D. *The Double Helix: A Personal Account of the Discovery of the Structure of DNA*. New York: Signet, 1968.

White, Lynn, Jr. *Medieval Technology and Social Change*. Oxford: Oxford Univ. Press, 1962.

Whittaker, Sir Edmund. *A History of the Theories of Aether and Electricity*. 2 vols. Vol. 1: *The Classical Theories*. 1910; rpt. New York: Harper Torchbooks, 1960.

Whyte, L. L. *The Unconscious Before Freud*. Garden City, N.Y.: Doubleday Anchor, 1960.

Will, Clifford M. "Gravitation Theory." *Scientific American* (November 1974), pp. 25–33.

Wilson, Bryan R., ed. *Rationality*. Oxford: Basil Blackwell, 1970.

Wilson, Colin. *The Occult: A History*. New York: Random House-Vintage, 1973.

Wilson, E. O. *Sociobiology: The New Synthesis*. Cambridge, Mass.: Harvard Univ. Press, 1975.

Young, Louise B., ed. *Exploring the Universe*. Oxford: Oxford Univ. Press, 1971.

Zaffron, Richard. "Identity, Subsumption, and Scientific Explanation." *The Journal of Philosophy* (1971), pp. 849–860.

INDEX

Acupuncture, 178–180
African scientific thought, 168
Alchemists, 153, 154
Alexander the Great, 145, 150
Alexandria, 145, 150
American Association for the Advancement of Science, 190
Analogy, 98
Antiquity, scientific thought in, 139–144, 191–192
Aquinas, Thomas, 154, 155, 159
Arab scientific thought, 153, 154
Aristotelian world view, 154, 156, 158
Aristotle, 141, 146, 148–151, 153, 154, 155
Assertions, 54–56
Assumptions, 112
  See also Presuppositions
Astrology, 132, 153, 192
  minimal hypotheses concerning, 194–196
Astronomy, 153
  in antiquity, 140–141
  Copernican, 156
  human uniqueness and, 4
Atomism, 165–166
Autonomic nervous system, 181–182
Aversive stimuli, 44

Babylonia, 140–141
Bacon, Roger, 154
Behavior patterns, natural selection of, 38–46

contingencies of survival, 43
emergence of variant behavior pattern, 44–45
established repertoire of behavior, 40
experimental behavior, 40–42
new problematic equilibrium, 45–46
reinforcement and aversion, 44
selective pressure, 42
species-specific behavior, 42, 43
Behavioristic approach, 40, 173–175, 189–190
Beliefs
  coherent, 61–64
  language's molding of, 54
  personal vs. scientific, 67–68
Biofeedback techniques, 181–183
Biology
  classification in, 85, 88–89
  explanations in, 99
  human uniqueness and, 4
  mechanical world view and, 174
  molecular, 88–89
  nomenclature in, 85
  relationships among describing, explaining, and identifying in, 91
Biot, Jean-Baptiste, 106, 107, 113, 117
Biota, 25, 27, 29
Bodily characteristics, natural selection of, 23–30
Body-mind (mind-matter) distinction, 67, 160–163, 172, 176

western psychology and medicine and, 176–178
Boethius, 153
Bohr, Niels, 166n
Brain, 8, 32
Buddhism, 151

Capra, Fritjof, 163, 168
Cargo cults, 192–193
Cartesian world view, 159–162, 167, 169, 172
  science of ourselves and, 171, 176, 177
  See also Body-mind distinction
Categorical presuppositions, 58–59, 110–111
Categories of thought, basic (species-specific), 50
Causal explanations, 71–73
Causality, 11, 17, 61
Change, Heraclitus' and Lao Tzu's views of, 146–147
Chemical elements
  evolution of, 88
  periodic table of, 87, 165
Chemistry
  control in laboratory situations, 95–97
  criteria for identifying substances in, 89
  in Middle Ages, 154
  Newtonian (mechanical) world view and, 164–165
  nomenclature and identification in, 84–85, 87–88
  relationships between describing, explaining and identifying in, 91
  theoretical explanation in, 92–95
Ch'hi (vital force), 151, 152, 178–179
China, 151–152, 154
Ch'in Empire, 152
Chi Ni (Chi Yen), 151

Chuang Tzu, 145, 151
Chu Hsi, 155, 157
Classical period, science in, 144–152, 155
Classifying, 84–88
  data vs. theory as basis of, 86–87
Clines, 47–48
  western and eastern, 134–136, 138
Coherent beliefs, 61–64
Color, 17–18, 36–37
Common sense experience, 11–12
  presuppositions of, 56–57, 61
Conceptual activity. *See* Thought
Confucianism, 145, 151
Conjecture, 112–115
Conscious control of bodily processes, 181–183
Consciousness, 172–174
  language and, 47
Consistency, 62–64, 115–116, 127
Contingencies
  of reinforcement, 44
  of survival, 43
Contradictions, 61
Control, experimental, 95–97
Convergence
  of eastern and western views of the self, 185–187
  of views of nature, 134–138
Copernicus, Nicolas, 155, 156, 158, 159
Copper smelting, 139–140
Counting, 12–13, 55, 70
Critical method (critical examination of hypotheses), 117–128, 131, 132, 189, 199–200
  ancient beliefs and, 191–193
  minimal hypotheses, criticism of, 193–197
  gravitation theories, current, 126–128
  two-fluid theory of electricity and, 117–125

Criteria for identification, 89, 90
Crucial experiments, 119-120

Darwin, Charles, 21-24, 25n, 26n, 28, 29, 176-177
Data, 81-84
    classification on the basis of, 86-87
    criteria and, 89
    definition of, 81
    theory and, 72, 81-82, 90, 97-98
Data-level hypotheses, 82
Definition(s)
    definition of, 89-90
    presuppositions compared with, 59-60
Deities, nature, 141-144, 197
Descartes, René, 155, 159-161
    See also Cartesian world view
Descriptions, 12-13, 85-86
    accurate, 76-78
    of color, 17-18
    data-level vs. theoretical, 97-98
    explanations and, 71-73
    general patterns of relationships among explanation, identification, and, 85-86
    See also Data
DNA-based taxonomy, 89
Dualism. See Body-mind distinction
Dolphins, 4, 8
Du Fay, Charles-François, 106, 107, 113, 116, 118, 119, 121, 123

Eastern cline, 134-136, 138
Eastern science. See Nature, eastern and western philosophies of
Eastern thought (in general), 134-136, 138
Einstein, Albert, 124-127, 166
Electricity, 102-125

fluid theories of, 104-125
    See also Hypotheses, selection of
Electromagnetic spectrum, 32, 33
Electromagnetism, 102-103, 108-109
Elements, basic, 148, 149
    See also Incorporeal agencies
Empedocles, 148
Endosomatic evolution, 39, 40
Equilibrium, natural selection and, 22, 25, 28-30
Ether, 124
Euler, Leonhard, 109
Event, notion of, 51
Evolution
    of chemical elements, 88
    endosomatic and exosomatic, 39, 40
    rate of, 25n, 28
    See also Natural selection
Existence presuppositions, 58
Exosomatic evolution, 39, 40
Experience, 7
    commonality of, 170-172
    common sense, 11-12
    presuppositions, 56-57, 61
Experimental behavior, 40-42
Experiments
    crucial, 119-120
    control in, 95-97
Explanation(s), 85-86
    causal, 71-73
    descriptions and, 71-73
    as human activity, 6, 17
    no uniquely correct, 13
    plausibility of hypotheses and, 64-65
    theoretical, 71-73, 91, 99, 142
Explication, 90
Extrasensory perception, 190-192
    minimal hypotheses concerning, 193-194
Eyes, human, 33-36

Feinberg, Gerald, 192
Formal presuppositions, 56–57, 60–61
Four-dimensional objects, 10–11
Franklin, Benjamin, 103, 105–107, 118–119, 121
Freud, Sigmund, 191

Galen, 137, 150, 153
Galileo, 61–62, 159, 160
Genghis Khan, 157
Geometry, 144, 146–148, 189
Goodfield, June, 120n, 168
Grammar, 47
Gravity, current theories of, 126–128
Greek thought, classical, 144–150, 153

Heaviness, 149
Heisenberg, Werner, 134, 167, 172
Heraclitus, 146, 147, 154
Herbal cures, 178
Hero, 145, 150
Hesiod, 142
Holism, 161–164, 186–187
Human beings
  as social animals, 46
  uniqueness of, 4
  See also Science of ourselves
Human experience. See Experience
Human self. See Self, the
Hydrogen generator, 92
Hypotheses, 54
  conjecturing, 112–115
  data-level, 82
  minimal, 193–197
  objective significance of, 120–121
  origin of, 113–114
  plausibility of, 64–68, 115–116
  selection of: debate over two-fluid theory of electricity as

example of, 110–125
  critical examination, 116–122
  emergence of thesis, 122–125
  established body of theory, 110–112
  plausibility, 115–116
  See also Natural selection, of scientific theories
  self-consistency of, 115–116
  theoretical, 82–83

Identifying, 69–70, 84–86
  criteria (keys) for, 89, 90
  general patterns of relationships involving, 85–86
  See also Nomenclature
Incorporeal agencies (subtile fluids), 102, 104–106, 137, 162
India, 151, 153, 157, 168, 183
"Inner states," 171, 181
Instinct, 43
Ionian nature philosophers, 137, 144

Jung, Carl, 191

Kant, Immanuel, 50n, 60n, 155
Kepler, Johannes, 155, 156, 159
Koestler, Arthur, 161, 172n, 202
Krafft, Karl Ernst, 196n

Language (linguistic behavior), 6, 8–9, 38
  beliefs as molded by, 54
  habits of thought and, 50–51, 54
  origin and development of, 47–52
  of science of ourselves, 170, 172
Language communities, differences among, 47–51
Lao Tzu, 145, 146–147
Lavoisier, Antoine, 165
Levitation hypotheses, 65–68

Light, visible, 32
Lightness, 149
Limiting presuppositions, 137, 188
Locke, John, 172*n*
Logic, 55
Logical positivism, 100

Mantra, 184
Mathematics, 144, 146–148, 153, 189
  Descartes and, 159–160
Maxwell, James Clerk, 104, 108, 109, 123
Maya (Hindu concept), 163
Mechanical world view, 162, 163, 188
  limits of, 164–168
  science of ourselves and, 169–175
  medicine and psychology and, 176–178, 180
  See also Body-mind distinction; Cartesian world view; Newtonian physics
Medicine, 153
  Cartesian (mechanistic) presuppositions and, 176–178, 180
  eastern approach to, 152, 178–181
Meditative techniques, 183–185
Mendelejeff, Dmitri Ivanovich, 87
Meteorology, 141–143
Middle Ages, scientific thought in, 152–154
Miller, Neal, 182
Mind
  conscious vs. unconscious, 172–173
  See also Body-mind distinction; Consciousness
Ming dynasty, 157
Minimal hypotheses, 193–197
Modeling, 98
Modern science, 139

See also Western physical sciences
Mohists, 151
Mongols, 157
Motion
  Aristotle's view of, 149–150
  Oresme's hypothesis about relative, 116–122
Muslim scientific thought, 153, 154, 155
Mutation, natural selection and, 26–29

Natural selection, 6, 99
  of behavior patterns. *See* Behavior patterns, natural selection of
  of bodily characteristics, 23–30
  biota, 25, 27, 29
  external challenges to mutants, 22, 27
  geographical location of mutant subpopulation, 27–29
  innovation (mutation), 22–29
  internal challenges to mutants, 22, 26–27
  new problematic equilibrium, 22, 28–30
  problematic equilibrium, 22, 25
  classification and theory of, 88
  of scientific theories, 128–132, 199
Nature, eastern and western philosophies of, 135–168
  in antiquity, 139–144
  from Aquinas to Newton, 154–161
  Cartesian. *See* Cartesian world view
  in classical period, 144–152
  holism (unitary view of reality), 161–164
  mechanical world view. *See* Mechanical world view

in Middle Ages, 152–154
Needham, Joseph, 162, 168
Nervous system, 6, 8
  autonomic, 181–182
Neutrality, 13
New Guinea, cargo cults of, 192–193
Newton, Sir Isaac, 102, 124–127, 155
  Descartes and, 160n, 161
Newtonian physics (or philosophy of nature), 102, 123–127, 164–167
  science of ourselves and, 169–175
  See also Mechanical world view
Nomenclature, 84–85, 90
  See also Classifying; Identifying
Northrop, F. S. C., 168

Objective significance of hypotheses, 120–121
Objectivity, 13, 170
  accuracy of descriptions and, 77–78
Observer-observed distinction, 171–172, 176
Occult, the, 189–190
Olmsted, Denison, 101–110, 116, 118–119, 121, 126, 127, 132, 133
One science, 130–132
Ontological presuppositions, 57–58, 110–112, 188
Oresme, Nicolas (of), 114–116, 122, 133, 155, 156
Organic unity of all reality, 162–163
  See also Holism

Parapsychology, 190–192
  See also Psychic phenomena
Pasteur, Louis, 114
Periodic table of the elements, 8ı, 165

Personalities of nature deities, 141–143
Personality types, 178
Philip of Macedonia, 150
Philosophies of nature. See Nature, philosophies of
Phlogiston, 120
Photons, 166
Phyletic change, 29, 137
Phylogenetic taxonomy, 84, 85
Physical objects
  common sense experience of, 11–12
  counting and describing, 12–13
Physical sciences, 102
  presuppositions of (in 1851), 110–112
  See also Science; Western science
Physics, 16, 101–102, 199–201
  Newtonian. See Newtonian physics
  quantum, 166–168
  subatomic, 166, 167
Planck, Max, 166
Planetary motion. See Astronomy
Plato, 147–148, 153, 154
Plausibility of hypotheses, 64–68, 115–116
Pluralism, 164
Pneuma, 148, 149, 152
Point of view, 13–14, 199
Poisson, S. D., 108, 109
Popper, Karl, 39n, 72, 119, 133, 199
Positivism, logical, 100
Possibility, theoretical, 63–66
Precedents, 64–65
Prediction, theoretical, 94–97
Presuppositions, 14–16, 56–61, 72
  categorical, 58–59, 110–111
  definitions compared to, 59–60
  existence, 58
  formal, 56–57, 60–61
  limiting, 137, 188
  ontological, 57–58, 110, 188

of science of ourselves, 170–172
as semantic items and assertions, 59–60
warranting, 189
Priestley, Joseph, 165
Proof, 80, 199
logical or mathematical, 189
Psychic phenomena, 66, 190–192
minimal hypotheses concerning, 193–197
Psychology, 16
behavioristic, 40, 173–175, 189–190
Cartesian presuppositions and, 176–178
human uniqueness and, 4–5
See also Parapsychology
Ptolemy, 150, 154
Pythagoras, 145, 146
Pythagorean school, 146, 147, 154, 156, 158

Quantum physics, 166–168

Re-description, 71–73, 81, 86, 94–96, 120, 122
Re-identification, 71–73, 128, 130
Reinforcement of behavior, 44
Relative motion, Oresme's hypothesis about, 116, 122
Relativity, general and special theories of, 124–127, 137
Religion, 197–198
ancient science and, 142
See also Deities, nature
Retina, 35, 37, 38
Roman Catholic Church, 154, 158, 159
Roman Empire, 145, 150
Rosenberg, Jay, 60n

Science (scientific activity)
eastern. *See* Eastern science
as group (communal) activity, 131
as human activity, 2–3, 31
as linguistic activity, 54
modern, 139
natural selection as applied to. *See* Natural selection, of scientific theories
proof and, 80
proper subject matter of, 189–191
as public activity, 54, 67–68, 79–80
religion and, 142, 197–198
western. *See* Western physical sciences
See also specific topics
Science of ourselves, 169–187
admissible data (scope) of, 169–176
Self, the
new conceptions of, 190–192
western vs. eastern views of, 176–187
Sensory apparatus, 6, 7, 31
limitations on conceptual activity and, 10–11
Shih Huang-ti, 152
Simpson, George, G., 52, 174
Skinner, B. F., 40, 41, 43, 49, 52
Social sciences, 118
See also Science of ourselves
Species, 25, 28
Species-specific traits, 36
Strawson, P. F., 50n, 52
Subtile fluids (incorporeal agencies), 102, 104–106, 137, 162
"Supernatural" phenomena, 190–191
Supposing, 54
See also Hypotheses
Survival, contingencies of, 43
Syntax, 47

Taoism, 147, 151, 152, 155, 157
Taxon, 85
Theology in Middle Ages, 153

Theoretical explanations, 71–73, 91–99
  contextual shifts and, 98
  theological explanations as, 142
Theoretical hypotheses, 82–83
Theoretical possibility, 63–66
Theoretical predictions, 94–97
Theory(-ies), 2–3, 17
  classifying on the basis of, 86–87
  data and, 72, 81–82, 90, 97–98
  established body of, 110–112
  new thesis' relationship to, 122–125
  natural selection of. *See* Natural selection, of scientific theories
Theses, 70, 83
Third-person perspective, 171–173
Thought (conceptual activity), 6, 7
  basic categories of, 50
  habits of, 50–51, 54
  limits of possible variation in, 10
Timaeus (Plato), 147–148, 153
Tin, 140
Toulmin, Stephen, 120n, 168, 187
Transcendental meditation, 184–185
Truth and falsity, 73–76

Unconscious activities of the mind, 173, 174
Unitary world view (holism), 161–164, 186–187
Unscientific views, 191–193
  See also Occult, the
Urban VIII, Pope, 61–62

Vedic traditions, 183, 184
Visible light, 32

Visual sense, 8, 10, 11, 32–37
  color discrimination, 36–37
  interpretation of visual stimuli, 34–35
"Vital force," 177
  See also Ch'hi
Vocabulary, data, 82

Wang Shou-jen, 155, 157–158, 167
Wei Po-Yang, 145, 152
Western cline, 134–136, 138
Western physical science, 1, 5, 9–10
  presupposition change in, 171–172
  proper subject matter of, 135–136, 189–191
  science of ourselves and, 171–175
  third-person perspective of, 171–173
  See also Nature, western and eastern philosophies of
Western thought (in general), 134, 135, 138
Wheatstone, Sir Charles, 103
Whyte, L. L., 172n
Will, Clifford M., 118n, 126–127, 132
Wilson, Edward Osborne, 43n

Xenophanes, 144

Yang and yin, 147, 151

Zeno, 147
Zoology, 82, 84, 89